RAND ARROYO CENTER

Evaluating the Army's Ability to Regenerate

History and Future Options

Shanthi Nataraj, M. Wade Markel, Jaime L. Hastings,
Eric V. Larson, Jill E. Luoto, Christopher E. Maerzluft,
Craig A. Myatt, Bruce R. Orvis, Christina Panis,
Michael H. Powell, Jose R. Rodriguez, Tiffany Tsai

Prepared for the United States Army

For more information on this publication, visit www.rand.org/t/RR1637

Library of Congress Cataloging-in-Publication Data is available for this publication.
ISBN: 978-0-8330-9663-0

Published by the RAND Corporation, Santa Monica, Calif.
© Copyright 2017 RAND Corporation
RAND® is a registered trademark.

Support RAND
Make a tax-deductible charitable contribution at
www.rand.org/giving/contribute

www.rand.org

Preface

This document reports results from a research project entitled, "Developing a Strategic Framework for Army Regeneration." The purpose of the project was to assess the Army's ability to regenerate active component end strength using a variety of accessions, retention, and force management policies.

This report presents a historical synthesis of the Army's efforts to expand during the decade following September 11, 2001. It identifies the various policy levers the Army can use to achieve its targets and conducts an empirical analysis of the limits on the Army's ability to expand under a variety of external conditions. It also identifies the larger policy implications of maintaining the capacity to expand as necessary.

This research was sponsored by the Deputy Chief of Staff G3/5/7, U.S. Army, and was conducted within the RAND Arroyo Center's Personnel, Training, and Health Program. RAND Arroyo Center, part of the RAND Corporation, is a federally funded research and development center sponsored by the United States Army.

The Project Unique Identification Code (PUIC) for the project that produced this document is HQD156908.

Contents

Preface iii
Figures and Tables vii
Summary ix
Acknowledgements xv

CHAPTER ONE
Introduction 1
Background 1
Purpose 2
How the Report Is Organized 3

CHAPTER TWO
Efforts to Expand the Army 5
Research Approach and Sources 5
Expanding Capacity 6
October 2001–December 2003: 9/11 Attacks to First End Strength Increase 7
January 2004–December 2006: First End Strength Increase to the Start of the
    Grow the Army Initiative 11
January 2007–December 2010: Duration of the Grow the Army Initiative 18
From January 2011 Forward: Drawing Down 23
Beyond Soldiers: Using Contractors to Augment Operational Capacity 23
Summary and Conclusion 25

CHAPTER THREE
Regeneration Scenarios 29
Estimating Required Operational Capacity 30
Conclusion 32

CHAPTER FOUR

Conceptual Framework and Policy Options for Regeneration...........................33

Policy Options for Regeneration 35

Summary...51

CHAPTER FIVE

Modeling Results 53

920K Scenario...53

980K Scenario...63

CHAPTER SIX

Conclusions and Implications for Preparation...........................69

Major Findings..70

Recommendations 77

Conclusion...81

APPENDIXES

A. Additional Modeling Results 83

B. Sensitivity of Results with Regeneration Wedges......................89

Abbreviations...95

References 97

Figures and Tables

Figures

2.1. Enlistment Incentives and Average Incentive Amounts.......................... 9
2.2. Army Reserve Component Members on Active Duty from September
2001 to June 2009 in Support of Operations Noble Eagle, Iraqi Freedom,
and Enduring Freedom 13
4.1. Conceptual Model of Active Component Regeneration 33
4.2. Conceptual Timeline of Active Component Regeneration 34
4.3. Average Continuation Rates for Each of Five Scenarios, by YOS............... 50
5.1. 920K Scenario: Estimated Enlisted Shortfall Using Different Policy Levers.... 54
5.2. 920K Scenario: Estimated Enlisted Shortfall Under Different Conditions....... 55
5.3. 920K Scenario: Estimated Enlisted Shortfall Under Different Conditions,
Greater Enlistment Eligibility.. 56
5.4. Immediate Demand.. 61
5.5. Gradual Buildup from Base Level of Demand................................. 62
5.6. 980K Scenario: Estimated Enlisted Shortfall Using Different Policy Levers.... 64
5.7. 980K Scenario: Estimated Enlisted Shortfall Under Different Conditions...... 64
5.8. 980K Scenario: Estimated Enlisted Shortfall Under Different Conditions,
Greater Enlistment Eligibility.. 65
A.1. 920K Scenario: Estimated Enlisted Shortfall Under Different Conditions...... 84
A.2. 980K Scenario: Estimated Enlisted Shortfall Under Different Conditions....... 85

Tables

S.1. Army End Strength Options Under Consideration ix
2.1. Department of Defense Contractors in Afghanistan and Iraq, 2007–2013..... 24
3.1. Army End Strength Options Considered..................................... 30
3.2. Estimating Additional Regular Army Strength Needed to Generate
Capacity Equivalent to 547K Regular Army Force............................. 31

4.1. Accession Scenarios and Inputs 39
4.2. Example of Regeneration Results for 450K, Average Recruiting Conditions, 5,821 OPRA Foxhole Recruiters, and Lesser Enlistment Eligibility Policies 41
4.3. Summary of Conditions and Assumptions Considered for Modeling Accessions, Retention, and Force Mix 52
5.1. 920K Scenario: Estimated NCO Experience 58
5.2. 920K Scenario: Estimated Incremental Costs of Recruiting and Retention 59
5.3. 920K Scenario: Estimated MOB:Dwell Ratio Assuming Immediate Increase in Demand 61
5.4. 920K Scenario: Estimated MOB:Dwell Ratio Assuming Linear Increase in Demand 63
5.5. 980K Scenario: Estimated NCO Experience 66
5.6. 980K Scenario: Estimated Incremental Costs of Recruiting and Retention 67
5.7. 980K Scenario: Estimated MOB:Dwell Ratio Assuming Immediate Increase in Demand 68
5.8. 980K Scenario: Estimated MOB:Dwell Ratio Assuming Linear Increase in Demand 68
6.1. Policy Options to Increase Army Capacity 71
6.2. Summary of Key Modeling Results 72
6.3. Accessing the Reserve Component for Operations 74
A.1. 920K Scenario: Summary of All Results 86
A.2. 980K Scenario: Summary of All Results 87
B.1. Regeneration Wedge Starting Conditions 90
B.2. 920K Scenario with Regeneration Wedge: Estimated NCO Experience 91
B.3. 980K Scenario with Regeneration Wedge: Estimated NCO Experience 93

Summary

Background and Purpose

In late 2011 and early 2012, the U.S. Army had largely ended its operations in Iraq and was reducing its commitments in Afghanistan. It was also beginning to substantially reduce the size of its forces in response to the Budget Control Act of 2011. According to the 2014 Army Posture Statement, under the fiscal year (FY) 2015 Budget request, the Army's active component (AC) end strength was to be reduced from approximately 490,000 to 450,000 soldiers, and the Army National Guard from 350,000 to 335,000, over the period from FY 2015 to FY 2017. The size of the U.S. Army Reserve was to be similar to its FY 2014 level of 195,000. These cuts would leave a Total Army of 980,000 (980K) soldiers. If further sequestration cuts were implemented, Total Army strength would decline to 920K, its smallest since World War II.

These cuts pose two problems for the Army. One is that they call into question the Army's ability to carry out its guidance from the U.S. Department of Defense (DoD). The other is its ability to restore itself to previous strength levels in a timely way, should the nation require it to do so. The research reported here examines the Army's ability to regenerate its AC end strength under two scenarios: one starting from a 420K AC (as part of a 920K Total Army) and one starting from a 450K AC (as part of a 980K Total Army). Table S.1 shows the distribution of personnel in each component under each scenario.

Table S.1
Army End Strength Options Under Consideration

Component	980K Option	920K Option
Regular Army	450,000	420,000
Army National Guard	335,000	315,000
Army Reserve	195,000	185,000
Total	980,000	920,000

SOURCES: McHugh and Odierno, 2014; McHugh and Odierno, 2015.

We focused on scenarios requiring rapid expansion to meet the demands of large-scale, protracted contingency operations of approximately the same scale as those in Iraq and Afghanistan since September 11, 2001. We did so for three reasons. First, these are the kind of contingency operations for which the 2012 Defense Strategic Guidance proposed regeneration—or "reversibility"—as a hedge (DoD, 2012). Second, using these scenarios enabled us to compare results with historical experience, making it possible to identify relevant variables and to reconsider assumptions. Third, in contrast to current planning scenarios used in DoD's support to strategic analysis, these examples are unclassified, allowing wider dissemination of the resulting analyses.

In developing these scenarios, we postulated that the Regular Army would have to expand to a degree that enabled the Army as a whole to produce the same amount of capacity it did at the end of fiscal year 2009 (what we refer to as a "550K AC"), when it reached its targets for expansion under the "Grow the Army" campaign. Obviously, doing so will require restoring the Regular Army to at least its 2009 baseline. Still more Regular Army strength will be needed to replace the reserve component (RC) strength that will also have been lost. Under the 980K scenario, the Regular Army would have to expand by almost 120,000 soldiers to produce the same operational capacity as the 2009 baseline. Under the 920K scenario, the Regular Army would have to expand by 160,000 soldiers. Both options are somewhat larger than the peak strength of the Army at the conclusion of the Grow the Army initiative, but both options must compensate for the substantial numbers of reservists no longer available under these options.

To examine the Army's capability to regenerate, we use a framework that integrates accessions, retention, and force management policies. We also quantify the costs and risks associated with each option and include suggestions about where the Army should make strategic investments and identify policy options to lay the foundation for future regeneration. This analysis applies to a long-term, large-scale counterinsurgency and stability operation. The Army probably cannot achieve a useful degree of expansion in time to meet sudden demands for additional combat power for short-notice, intense operations, such as potential contingencies in the Baltic States or in Korea.

While several prior studies have examined regeneration options, these studies have tended to focus on individual regeneration challenges or to consider fairly limited surge scenarios. This research builds on prior work by creating new knowledge along four dimensions. Specifically, it

- undertakes the first historical synthesis of the Army's efforts to expand during the decade following the terrorist attacks that occurred on 9/11
- considers regeneration holistically, identifying multiple policy levers the Army can use to achieve its targets
- conducts an empirical analysis of the limits on the Army's ability to expand under a variety of external conditions

- identifies the larger policy implications of maintaining the capacity to expand as necessary.

Approach

We developed a conceptual model of regeneration that reaches target end strength after a number of years by applying various policy levers to affect the flow of soldiers into and out of the force. The Army can affect these flows using different accession and retention policies, including how it allocates funding and personnel to the recruiting and retention efforts, and can also influence the internal composition of the Army by adjusting promotion and retention policies. How many recruiters it has, how much it spends on advertising and incentives, and what enlistment eligibility criteria it imposes have important effects on recruiting. Similarly, the rate of promotion and the size and number of reenlistment bonuses it offers affect retention rates. Because the Army does not recruit in a vacuum, we have also posited different external environments that affect the Army's ability to attract recruits from the civilian population. These environments include such things as the job market (a bad one is generally good for Army recruiting) and how the civilian population might view a given conflict. The analysis also took into account the extent to which the RC could be mobilized to reach targets.

We entered all these factors into a combined modeling framework to assess the Army's regeneration ability under different combinations of policies and external conditions. These included varying the number of recruiters, selectively awarding enlistment and reenlistment bonuses, allowing different eligibility benchmarks, and considering the effects of different unemployment rates. We also estimated the costs involved with different policy combinations and their effects on the frequency of RC deployments.

What We Found

Current Policy Levers Will Probably Suffice to Enable Regeneration

Our analysis indicates that the current suite of policies the Army and DoD have at their disposal is likely adequate to expand the force to provide the capacity associated with a 550K AC, starting with either the 980K or 920K Total Army. While our analyses did not uncover any constraints that would make such regeneration infeasible, they do suggest that the effort would carry a number of risks, particularly when expanding from a Total Army of 920K. The most potentially critical risk revolves around the fact that, while the Regular Army is expanding, the Army as a whole will still need to meet operational demands. Thus, the Army will have to draw upon its RC to an unprecedented extent to sustain high levels of operational commitment until it accomplishes regeneration. The required rotation of the RC that we estimated should be feasible under the current authorities. However, RC forces may require mobilization periods exceeding the one-year limit for involuntary mobilization in current DoD policy to achieve the standard of proficiency needed to replace Regular Army forces for a useful length of time at acceptable risk. In addition, although Army National Guard leadership has expressed a willingness to operate at a tempo of 1:2, post 9/11 experience suggests that doing so may erode congressional or public support for sustained use of the RC. Achieving the target AC strength, particularly when starting from a Total Army of 920K, will also likely require expanding enlistment eligibility, thus lowering the quality of the average recruit. The Army will also need to be able to leverage extensive contract support throughout the duration of the conflict. We also note that many of these policy options—notably increasing authorized end strength and various options for mobilizing the RC—rely on decisionmakers outside the Army.

External Conditions Matter a Lot

As noted, our analysis indicates that extant policy levers will enable the regeneration of the force. However, that analysis drew on conditions present in the past. Whether they will work in the future depends on external conditions at that time; the willingness of the Army, DoD, the President, and Congress to use existing policy tools; and the willingness of the American public to respond to their use. To cite only one issue, if Congress and the people do not support involvement in the conflict, Congress might not want to raise the end strength or be willing to allow deployment of the RC at a high frequency.

Regeneration Would Stress the Reserve Component, Especially When Starting from 920K

All regeneration scenarios we considered would require the RC to rotate at a cumulative mobilization-to-dwell ratio of less than 1:3 over a number of years. While the RC can sustain this level of deployment from a force structure perspective, it would place the RC under substantial stress. This might make it difficult to muster or sustain congressional and public support. Although demand for RC forces would decline as the AC expanded, they would still be deployed at a cumulative ratio below the 1:3 threshold into the sixth year.

What We Recommend

Simply put, regeneration to the levels described above is feasible—in theory. However, there are two important "ifs." *If* the Army has prepared for that contingency and *if* DoD officials make and implement challenging and unpalatable decisions early, regeneration from either a 980K or a 920K Total Army may be accomplished. That said, both scenarios entail risk. For example, the AC cannot be regenerated overnight, so the

Army will need to rely heavily on the RC for several years to bridge any gaps between required and available AC operational capacity.

The following recommendations address actions that lie largely within the remits of DoD and the Army. These actions require public support to succeed. Political leaders—especially the President—must be prepared to expend political capital to generate and sustain the necessary level of public support and create a context in which measures to expand the Army can succeed.

Develop Planning Scenarios Requiring Regeneration

The Army needs to develop and resource specific capabilities to enable regeneration. These capabilities include such things as recruiters, institutional trainers, infrastructure, equipment, and leaders in units. The Army must determine how many and what kinds of assets it needs either to maintain in the inventory or to produce during regeneration. Thus, it should explore a range of scenarios to make informed decisions about what specific requirements it would have to meet and to determine which of those exist now and which need to be developed.

Assess Alternative Ways to Posture the Army for Regeneration

This analysis focused on how to generate the recruits necessary to staff an expanding Army. Less thought has been given to receiving and managing such an expansion. In the past, the Army has considered different approaches to expansion—establishing cadre formations, undermanning units during peacetime to be filled out in war, drawing on manpower from its generating force, and relying on RC units to fill critical gaps in larger units. The Army should explore which of these approaches, or what combination thereof, best positions it to expand rapidly.

Prepare the Reserve Components for Rapid and High-Frequency Deployment

Our findings make it clear that the Army will need to rely heavily on the RC for a substantial length of time while the AC is being regenerated. The rotation tempo we estimated should be feasible under the current authorities, and Army National Guard leadership has expressed a commitment to operate at a 1:2 mobilization-to-dwell ratio. However, during operations Enduring Freedom and Iraqi Freedom, the Army encountered congressional resistance when it started deploying the RC at or above a 1:3 mobilization-to-dwell ratio. In addition, current DoD policy may need to be revised to allow involuntary RC deployment for more than one year and at the required tempo.

Thus, it will be critical to prepare all relevant decisionmakers today for the possibility of rapid, high-frequency deployment in the event of a conflict that requires regeneration. Failure to prepare all stakeholders for these commitments risks disruption at the time of crisis. Such preparation may involve changing policies that limit RC employment, such as the one that limits exceeding one year of mobilization.

Maintain Certain Critical Skills in the Army Today to Reduce the Stress on the Army During Regeneration

This analysis has essentially treated soldiers as generic. But some soldiers required to build up the force will need specialized skills—pilots, medical staff, and special operations forces—and these may be impossible to build in a short time. Similarly, it will be necessary to have midgrade officers and noncommissioned officers to expand the force. Such individuals will be especially important if the Army must expand from the 920K level and have to accept somewhat lower-quality recruits to reach the desired end strength. Maintaining a wedge of additional midgrade leaders, along with a sufficient number of individuals in military occupational specialties with long lead times for training or skills development, could reduce the risk associated with a potential regeneration; however, the benefits of the reduction in risk would have to be compared with the cost of maintaining the wedge during peacetime.

Maintain Army Capacity for Contingency Contracting

The Army has depended on a large contracting force in recent conflicts. Assuming a relatively permissive threat environment, that dependence will likely continue in future conflicts. But such contracts need to be managed carefully to avoid the waste and abuse that occurred in prior conflicts. As the Army reduces its end strength, the natural tendency will be to reduce the acquisition workforce to levels commensurate with the supported force. The Army should resist that tendency. As Army operating forces decrease, the need to contract support and sustainment capacity may well increase and certainly will not decrease.

Develop Contingency Plans

Our analysis has not identified any definitive limit to the Army's ability to regenerate at the speed and on the scale described in this analysis but has indicated that these efforts are fraught with risk. The maximum accessions the Army has achieved since 2001 were around 80,000 a year. As discussed in the text, that was the Army's objective at the time, so we cannot assume it constitutes a limit. It may be a warning, however. For that reason, as the Army plans for rapid expansion of the Regular Army, it should also develop contingency plans in case that expansion falters. The contingency plans will almost certainly hinge on a much higher degree of mobilization of the Army's RC.

Decide Early

This analysis assumes that the decision to regenerate rapidly would be made at the start of the conflict and that all policy levers would be in place by the end of the first year of the conflict to deliver the first meaningful increment of capability by the third year. If the decision lags, the time lines described here would also. If decisions can be made even more quickly than we assume, increased recruiting and retention may be possible even during the first year. For that reason, a decision to go to war should be a decision to expand the Army.

Acknowledgements

We thank Daniel Klippstein, Deputy Director for Plans and Policy, Headquarters, Department of the Army, Strategy, Plans and Policy, for sponsoring this research. We are grateful to COL Jeff Hannon, Chief, Strategy, Concepts and Doctrine Division for his leadership in developing the study's overall framework, and to MAJ Seth Chappell and Walter (Tony) Vanderbeek for their guidance and input throughout the study. We also thank MAJ Bryan Rozman for his helpful comments on our work in progress.

We are indebted to BG Michael J. Meese (U.S. Army, ret.) and Jennie Wenger, who provided valuable reviews. We appreciate the support and analytical insights offered by Michael Hansen, Director, Personnel, Training, and Health Program, RAND Arroyo Center. We also thank Steven Garber, who originally developed the accessions model, and Henry (Chip) Leonard, who originally developed the inventory model we use in our analyses and who provided an informal review of this report.

CHAPTER ONE

Introduction

Background

The end of 2011 and the beginning of 2012 marked an inflection point for the United States military (U.S. Department of Defense [DoD], 2012, p. 1). The United States was reducing its forces in the Middle East: The war in Iraq had ended, and military officials hoped to reduce American involvement in Afghanistan, handing off major responsibilities to the Afghan government. In addition, the federal government, including DoD, was facing significant budget cuts that had been imposed by the Budget Control Act of 2011 (DoD, 2012; Public Law 112-25, 2011). To meet these requirements, all the defense services and agencies have been forced to make difficult decisions, but the Army in particular—given its large number of personnel—was called on to substantially reduce its force size.

The 2014 Army Posture Statement states that, under the fiscal year (FY) 2015 budget request, the Army planned to draw down its active component (AC) forces from an end strength of 508,000 soldiers in FY 2014 to 450,000 soldiers by FY 2017 (McHugh and Odierno, 2014). At the same time, the Army National Guard (ARNG) would be reduced from approximately 354,000 soldiers at the end of FY 2014 to 335,000 soldiers, and the U.S. Army Reserve (USAR) would remain close to its FY 2014 level of 195,000 soldiers (for a Total Army of 980,000 [980K]).[1] If further cuts are required under sequestration, the projected end strength would fall to 420,000 for the AC; 315,000 for the ARNG; and 185,000 for the USAR by FY 2019 (for a Total Army of 920K). If these reductions take place, the proposed size of the AC would be the lowest since World War II (Shanker and Cooper, 2014) and would be 120,000–150,000 below its recent peak of approximately 566,000 in FY 2010.[2]

The cuts directed are of such magnitude that they threaten both the Army's ability to carry out the defense guidance and to reverse course quickly, should that be required. The DoD's 2012 Defense Strategic Guidance (DSG) notes that "U.S. forces

---

[1] Estimated end strengths for FY 2014 are from the Office of the Assistant Secretary of the Army (Financial Management and Comptroller) (FM&C), 2015a, p. 9; FM&C, 2015b, p. 7; FM&C, 2015c, p. 7.

[2] The end strength for 2010 is from FM&C, 2011, p. 10.

will no longer be sized to conduct large-scale, prolonged stability operations" but that DoD must "protect its ability to regenerate capabilities that might be needed to meet future, unforeseen demands" and apply "the concept of reversibility" when making decisions about current investments (DoD, 2012). The 2014 Army Posture Statement notes that "significant risk" would be associated with executing the 2012 DSG, given an AC of 450,000 and associated reserve component (RC) levels, and that the DSG could not be executed given continued sequestration cuts; moreover, with an AC end strength of 420,000, the "reduction in our institutional base will make reversibility more difficult" (McHugh and Odierno, 2014, pp. 2–3).

Purpose

This report examines the Army's ability to regenerate AC end strength in two scenarios: starting from 420K (as part of a 920K Total Army) or from 450K (as part of a 980K Total Army) to the capacity generated by the size of the Army as seen at the end of the last conflict within five years. We refer to the target end strength as "550K AC," but, as we describe in detail in Chapter Three, the critical measure is the number of deployable troops provided not only by the AC but also by the associated RC.

To assess the Army's capability to meet regeneration targets, we develop a strategic framework that integrates accession, retention, and force management policies to identify various options. We consider and, to the extent possible, quantify the costs and risks associated with each policy option, including suggestions for where the Army should make strategic investments and policy changes today to lay the groundwork for future regeneration. This analysis applies to a long-term, large-scale counterinsurgency and stability operation. The Army probably cannot achieve a useful degree of expansion in time to meet sudden demands for additional combat power for short-notice, intense operations, such as potential contingencies in the Baltic States or in Korea.

Prior studies of regeneration have tended to focus on individual regeneration challenges or to consider fairly limited surge scenarios. For example, Orvis et al., 2016, models the potential for using accession policies to grow three additional brigade combat teams (BCTs) over one to two years. Other studies (Horowitz et al., 2012; Klimas et al., 2014) have focused on the optimal AC/RC mix to support surge operations.

This research creates new knowledge along four dimensions by

- undertaking the first historical synthesis of the Army's efforts to expand during the decade following the terrorist attacks that occurred on September 11, 2001
- considering regeneration holistically, identifying multiple policy levers the Army can use to achieve its targets

- conducting an empirical analysis of the limits on the Army's ability to expand under a variety of external conditions
- identifying the larger policy implications of maintaining the capacity to expand as necessary.

How the Report Is Organized

Chapter Two examines efforts to expand the Army over the past 15 years, particularly the "Grow the Army" initiative undertaken in 2007–2010. Chapter Three lays out the two different regeneration scenarios we considered, and Chapter Four summarizes the methodology for analyzing how accession, retention, and force mix policies could be used to meet regeneration targets under a variety of conditions. Chapter Five presents key results, and Chapter Six discusses implications for planning and preparation. Two appendixes supply additional model results and discuss the sensitivity of our results.

CHAPTER TWO

Efforts to Expand the Army

This chapter provides insights about the Army's ability to regenerate in the future by drawing from the experience of the 15 years since September 11, 2001, with a particular focus on the factors that limit the speed and scale of regeneration. Although the chapter touches on limits imposed by policy and strategy, it primarily explores the practical limits imposed by the force's ability to handle the strain of providing the required capacity; Americans' willingness to serve in the military, given the conditions that obtained at the time; and the resources available to induce them to serve.

Research Approach and Sources

This chapter largely represents a synthesis of the secondary literature and available data. The data are drawn from such diverse sources as Army budget justification documents, Congressional Research Service reports, the Congressional Budget Office (CBO), the Government Accountability Office (GAO), and other sources. Qualitative data, used mostly to illustrate how key stakeholders viewed conditions, objectives, and available options at the time, are taken from the aforementioned sources, secondary works about the war, and congressional testimony. Note that no single, consistent source of publicly available data describes the Army's deployments and capacity. These data are absolutely essential, however, for quantifying the limits and the possibilities for future expansion.

In our assessment of the Army's potential capacity to recruit and retain soldiers, we focus on several factors that previous analyses had identified as being important: the state of the economy, the intensity and success of ongoing military operations, eligibility criteria, and recruiting resources.[1] Thus, this chapter is more illustrative than exploratory; it reveals no new factors and indicates no new theoretical framework. It does, however, illustrate how these factors have played out recently in the real world, as the Army undertook unprecedented yet critical initiatives to sustain and increase capacity under conditions of great stress and uncertainty.

---

[1] Recruiting resources include the number of recruiters, bonuses, and advertising (Kapp, 2002). Also see the discussion of casualties in Asch, Heaton, et al., 2010. We chose to interpret casualties as an indicator of how the war was going at any given time.

Expanding Capacity

In response to both changing conditions in theater and domestic concerns, the Army employed several different but complementary approaches to expand its capacity. First, it reorganized or rebalanced so that more of the forces it had on hand were useful in ongoing conflicts. For instance, it converted air defense artillery and long-range fires units, neither of which was especially important in counterinsurgency operations, into military police and civil affairs units. Second, it concentrated available resources on the fight at hand, drawing military manpower from the generating force to create additional operational units. Third, it drew on RC capacity, mobilizing its members at a MOB:Dwell ratio of one mobilization year to approximately three years in some sort of "dwell" status,[2] beginning with the 2003 invasion of Iraq and continuing through the 2009 start of the Afghan "surge." Finally, the Army sought to expand its AC, starting in 2004 and continuing through the end of the Grow the Army initiative in 2010. Throughout the entire 15-year period, the Army relied heavily on contractors to provide logistic support and elementary security. All these measures were necessary; none were sufficient, in and of themselves.

The Army's historical experience suggests that its ability to expand is highly sensitive to current conditions, including the state of the economy. When economic conditions are poor, recruiting conditions are favorable. For example, after the 2008 financial crisis, the Army increased its AC's size by about 47,000 soldiers in just three years after the establishment of the objective, instead of the six years for which it had planned. The Army also easily met its objectives for both quantity and quality and had extensive waiting lists of individuals who wanted to join as soon as possible. Conversely, in 2006, unemployment was a mere 4.6 percent nationally, and the Army was able to meet its accession objective only by both vastly increasing recruiting and retention resources and relaxing eligibility standards.

Another potential limit to the Army's ability to expand is the public perception of the associated war or conflict. For example, the Army's experience in the FY 2006–2008 period was affected both by a stronger economy and lower support for the Iraq and Afghan wars, making regeneration difficult. First, when the Army started deploying the RC at or above the 1:3 MOB:Dwell ratio, it started encountering congressional resistance. Second, the Army almost never reduced the AC's BOG:Dwell ratio below 1:1 and, even then, not for very long. Third, during these stronger economic times, the Army recruited only about 80,000 soldiers per year and that only when it relaxed eligibility standards. However, taken alone, these potential external limits are not indicative of constraints on Army regeneration. During the 15-year period examined, the

---

[2] Joint Publication 1-02 (2015) defines *dwell time* as the period a soldier spends between tours of involuntary active duty. Conversely, the time a soldier spends deployed overseas in a combat environment is called *boots-on-ground* (BOG) time for an AC soldier and *mobilization* (MOB) time for a reservist. The current BOG:Dwell goal for AC soldiers is 1:2, and the current MOB:Dwell goal for RC soldiers is 1:5.

Army and DoD were ambivalent about expansion, had not anticipated the crisis they faced, and consequently failed to deploy adequate resources effectively. Future Army leaders may face similar inefficiencies and similar limits. Despite potential external and internal constraints, better performance is not out of the question if the Army prepares effectively and deploys the necessary resources in a timely and effective fashion.

To better evaluate the Army's experience since 9/11, we have organized the chapter by key periods we determined reflected different phases in the Army regeneration process.

October 2001–December 2003: 9/11 Attacks to First End Strength Increase

This period was defined by both policymaker assumptions about appropriate troop levels and conflict duration and a favorable climate for expanding Army capacity. Before Operation Enduring Freedom (OEF) began in October 2001, there were few, if any, preexisting assumptions about ground force requirements in Afghanistan. A limited force of special operations forces, Central Intelligence Agency paramilitary personnel, and indigenous Afghan forces seemingly led to a quick and decisive victory over the Taliban. Shortly thereafter, in the early stages of Operation Iraqi Freedom (OIF) in 2003, U.S. forces swiftly secured Baghdad, and the Ba'athist government collapsed. These rapid successes reinforced Office of the Secretary of Defense assumptions that both conflicts would require modest U.S. troop levels for short-term engagements with control quickly being turned over to indigenous forces.

However, following the overthrow of Saddam Hussein's regime, senior DoD civilian and military leaders differed over the military's role in Iraq's postcombat environment and the troops necessary to fulfill that role, with DoD civilians taking a more optimistic view of force requirements and operation duration.[3] Even so, policymaker assumptions leaned toward more modest troop levels and shorter-term engagement.

In tandem with these assumptions, the Army faced relatively favorable conditions for expanding its capacity. At the beginning of the decade, the Army often struggled to meet its accession and retention goals in the face of strong economic growth, high employment, and other factors (Kapp, 2002). However, the mild 2001 recession led to slight increases in unemployment, reducing private-sector competition for

---

[3] During this period, Iraq war plans went through several iterations, each with differing troop levels and time frames. Although this issue is debated in secondary sources both contemporaneous with and following the Iraq war, evidence suggests that Secretary of Defense Donald Rumsfeld sought to shape Iraq war plans at most stages of the planning process. The war plans Rumsfeld proposed emphasized minimizing forces and shortening deployments, which observers interpreted as a crystallization of his vision of a smaller, increasingly streamlined military. Contrarily, evidence also suggests some commanders sought additional forces and longer time lines to ensure sufficient forces and capabilities to initiate and maintain postcombat operations.

labor.[4] Unemployment rose from an average of 4.6 percent in FY 2001 to an average of 6.0 percent in FY 2003; in 2003, the unemployment rate peaked at 6.5 percent in June (U.S. Bureau of Labor Statistics [BLS], 2015). Meanwhile, the American public supported both military efforts in Afghanistan and prospective operations in Iraq. A Gallup poll found that, as of December 2003, 71 percent of Americans approved of "U.S. military action in Afghanistan" ("Afghanistan," undated). Another December 2003 Gallup poll found that 64 percent of Americans supported the Iraq invasion and that 65 percent approved "of the way the U.S. has handled the situation with Iraq since the major fighting ended in April 2003" ("Iraq," undated). Propensity to enlist in November 2003 exceeded even that achieved in November 2001 in the immediate aftermath of the 9/11 attacks.[5] Approval for military operations may have remained high because the costs in blood and treasure remained low during this period. By the end of FY 2003, U.S. forces had suffered only 341 deaths from hostile action (Defense Manpower Data Center [DMDC], 2011). Casualties in these conflicts remained low. In short, economic headwinds against recruiting abated somewhat, and the conflicts for which the Army had to generate manpower were relatively popular.

The Army did not rely solely on a favorable climate. In what would become a common approach to meeting demand, both DoD and Congress authorized end strength increases for the Army AC.[6] On September 14, 2001, President George W. Bush signed Executive Order 13223 (Bush, 2001), declaring a national emergency. As part of Executive Order 13223, President Bush invoked his statutory authority under a state of emergency to suspend certain constraints on personnel management and delegated that authority to the Secretary of Defense. He also waived restrictions on end strength.[7] As will be seen, DoD made sparing use of this authority. Additionally, President Bush invoked his partial mobilization authority under 10 USC 12302 to tap into the RC generating force. Under these authorities, in November 2003, Congress authorized a modest permanent AC end strength increase of 2,400 for the Regular Army. Two months later, Secretary of Defense Rumsfeld used his delegated authority

---

[4] According to economists, the 2001 recession began in March 2001 for many reasons, including the bursting of the 1990s technology bubble, and continued through November 2001 (Kliesen, 2003; Langdon, McMenamin, and Krolik, 2002).

[5] See Commission on the National Guard and the Reserves, 2008, Table 1.8, p. 77.

[6] In passing the National Defense Authorization Act for Fiscal Year 2004 (Public Law 108-136) in November 2003, Congress authorized a modest permanent end strength increase of 2,400 for the Army AC. Two months later, Secretary of Defense Rumsfeld authorized an additional temporary end strength increase of 30,000 for the Army AC.

[7] Section 4 of Executive Order 13223 permits the Secretary of Defense to exercise presidential authority under 10 U.S. Code (USC) 123 (suspension of laws related to promotion, involuntary retirement, or separation of military commissioned officers), 123a (waiver of annual statutory end strength caps for military personnel), 527 (authority to suspend authorized end strengths and distributions in grade for general officers, flag officers, and commissioned officers above lieutenant colonel), and 12006 (authority to suspend authorized strengths and distribution in grade for commissioned officers and reserve general and flag officers in active status).

to establish a temporary end strength increase of 30,000 for the Regular Army. The RC was mobilized for operations both inside and outside the continental United States; by 2003, RC usage rates had exceeded those during the 1991 Gulf War.[8] Both President Bush and President Barack Obama annually renewed the provisions of Executive Order 13223, enabling the Army's continued reliance on end strength waivers and RC mobilization.

The Army also increased some recruiting and retention resources slightly. The number of recruiters for the Regular Army rose from 5,156 in FY 2001 to 6,367 in FY 2002 but fell back to 6,078 in FY 2003 (CBO, 2006, p. 9). The number of enlistment incentives for contracts signed with non–prior-service recruits rose slightly between FY 2001 and FY 2002 but fell in FY 2003 (Figure 2.1).

In this climate and with these measures, the Regular Army was able to exceed not only its recruitment goals in terms of quantity and quality but also its retention

Figure 2.1
Enlistment Incentives and Average Incentive Amounts

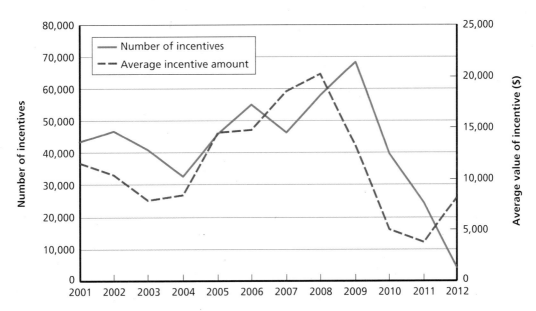

SOURCE: Authors' calculations based on data from Regular Army (RA) Analyst from FYs 2001–2012.
NOTES: *Number of incentives* includes the total number of non–prior service contracts that included any type of incentive, including cash enlistment incentives and educational incentives. *Average value of incentive* is the mean total value of these incentives among contracts that included an incentive. Educational incentives are valued using the actuarial value. All values are given in 2012 dollars based on the consumer price inflation index from BLS, 2015.
RAND *RR1637-2.1*

---

[8] Between 1992 and 2001, RC forces performed an average of 40 days of duty per year, including training days and support for operational missions; in 2002, that number exceeded 80 days on average; and in 2003, it exceeded 120 days (GAO, 2004).

goals. In FY 2003, the Regular Army achieved an end strength of 499,000, including overstrength and personnel affected by stop loss against an authorized end strength of 480,000. In numbers, it recruited 74,132 soldiers against a goal of 73,800. In terms of quality, the recruits exceeded DoD goals of 90 percent of them being high school diploma graduates and 60 percent of them having above-average scores on the Armed Forces Qualification Test (AFQT): 94 percent of Army recruits had high school diplomas, and 71 percent scored above average on the AFQT (FM&C, 2004, p. 5). These results were a slight improvement—at least in terms of recruit quality—over those for FY 2001. In that year, the Army met its goal for having 90 percent of its recruits have high school diplomas, and 63 percent of the recruits scored above average on the AFQT (Kapp, 2002, Table 2, p. 4). The Regular Army exceeded its retention goals in FY 2003 by slightly higher margins than it had in FY 2001 as well.[9] Thus, the Regular Army was able to achieve a modest improvement over its precrisis recruiting performance under favorable conditions, enabling it to maintain approximately the same size and quality it had achieved by the beginning of the decade.

A less-popular force management option during this time was stop loss.[10] Stop loss has existed since 1984 and has been used in previous conflicts. It includes skill-based stop loss, intended to retain personnel with critical skills; unit-based stop loss, intended to maintain unit strength and integrity; and stop movement, which precludes the loss of unit personnel because of reassignment orders (Henning, 2009, p. 2). All services used stop loss during OEF and OIF, but the Army relied the most on the practice. On September 24, 2001, Secretary of Defense Rumsfeld delegated his stop-loss authority to the heads of the military departments, making it easier for them to use this tool in the ways that best suited an individual service's needs.

Although the Army grew modestly during this period, this growth took place in an environment of certain DoD assumptions and stakeholder resistance to other available policy options for Army generation. As mentioned previously, although recruiting and retention conditions were favorable, DoD felt no need to significantly enlarge the Army during this period because of assumptions that force requirements would be modest. Additionally, a return to conscription, also known as the "draft" under the Selective Service Act, was undesirable and not seriously considered by civilian or military leaders. Secretary of Defense Rumsfeld and other senior DoD leaders preferred the All-Volunteer Force to conscription and would not consider the latter as an option. In 2003, Representatives Charles Rangel of New York and John Conyers of Michigan introduced legisla-

---

[9] Compare the FY 2001 reenlistment rates in Kapp, 2002, Table 11, p. 23, with those recorded in the table entitled "Performance Metric: Active Enlisted Retention Goal" in Office of the Under Secretary of Defense (Comptroller) (OUSD[C]), 2003, p. 374.

[10] *Stop loss* is a DoD force-management tool permitting the services to temporarily halt all voluntary separations and retirements during wars and national emergencies. The practice requires enlisted service members to stay in service beyond their contracted separation date. It does not apply to officers because they do not have established separation dates. Authority for stop loss is codified in 10 USC 12305. See also Henning, 2009.

tion to reinstate the draft, but it died in the House as a result of an overwhelming lack of support, with 402 representatives voting against the measure. Although Gallup and Pew Research Center polls show a slight uptick in public support for the draft immediately following 9/11, that support peaked at 20 percent and quickly waned (see, for example, Carlson, 2003; Carlson, 2004; Jones, 2007; Pew Research Center, 2011).

January 2004–December 2006: First End Strength Increase to the Start of the Grow the Army Initiative

During this period, demand abated somewhat from its 2003 peak in terms of absolute numbers, but stress on the force increased. It became apparent, however, that even with a slight drop, demand would continue at elevated levels for quite some time. Army strength in Iraq rose to approximately 150,000 in September 2003 and hovered between 100,000 and 132,000 thereafter. Meanwhile, military strength in Afghanistan rose from about an average of 10,400 in FY 2003 to an average of 20,400 of FY 2006, of which 17,100 were Army (see DMDC, undated).[11] The Army also reached the limits of its capacity. Throughout this period, the Army deployed Regular Army soldiers at a ratio of 1:1, far lower than the established BOG:Dwell goal of one year deployed to two years based at home station (Bonds, Baiocchi, and McDonald, 2010, p. 54). Operational demands were quite high relative to the Army's available operational capacity.

DoD assumptions regarding the time and troop levels needed also began to shift. For example, the February 2006 *Quadrennial Defense Review Report* opened with the statement: "The United States is a nation engaged in what will be a long war" (DoD, 2006, p. v). This signaled that Secretary of Defense Rumsfeld, and DoD, recognized the wars in Afghanistan and Iraq and the "Global War on Terror" likely would not be concluded in the near future. The 2006 *Quadrennial Defense Review Report* (DoD, 2006) also endorsed a permanent Army force structure of 70 BCTs, a number later deemed too small in a fall 2006 revised analysis of global strategic BCT requirements (see U.S. House of Representatives, 2007a, p. 4).[12]

In response to the actual and anticipated increases in operational demand and concomitant stress on the force, the Army undertook a number of initiatives. A modular transformation was the centerpiece of its effort. The modular transformation aimed to render Army forces more employable by restructuring the Army's division-based force to one organized around BCTs. The Army increased the number of maneuver brigades, in part by making them smaller; each BCT had two maneuver battalions, in contrast to the three maneuver battalion brigades that they replaced. Closely allied to

---

[11] See also Belasco, 2009, Table 1, p. 9.

[12] In December 2006, the House Armed Services Committee released its own defense review, raising questions about the adequacy of 70 Army BCTs (see House Armed Services Committee, 2006, pp. 71–72).

the modular transformation was rebalancing, in which the Army increased the number of units for which demand was high relative to the available supply—infantry, military police, civil affairs, and intelligence—at the expense of unit types for which demand was anticipated to be low—air defense artillery and long-range fires.[13] In testimony before the Senate Armed Services Committee, then–Army Chief of Staff GEN Peter J. Schoomaker asserted that this modular transformation would "increase the combat power of the active component by 30 percent as well as the size of the Army's overall pool of available forces by 60 percent" (Schoomaker, 2005, p. 22).

Although modular transformation and rebalancing helped the Army grow, similar opportunities to increase capacity through reorganization may not be present in future circumstances. Thus, we focus on measures undertaken to sustain capacity while the modular transformation and rebalancing were under way. These enabling measures included drawing on the generating force's military manpower, relying on RC combat power, and temporarily increasing the end strength by 30,000 Army soldiers. We discuss each of these enabling efforts in the succeeding paragraphs.

Drawing on the Reserve Components

The Army drew on the generating force's military manpower both to increase the size of operating forces and to meet immediate operational requirements for additional staff officers, trainers, and similar functions. The former objective was to be accomplished by converting military positions in the generating force to civilian positions, with a planned conversion of 11,000 Regular Army billets over time (Harvey and Schoomaker, 2005, p. 18). By the end of FY 2006, the Army had converted 9,600 such billets.[14] The Army also drew on the generating force to fill billets in contingency headquarters, such as when Multinational Force–Iraq hastily assembled training teams under the direction of the newly established Multinational Security Transition Command–Iraq.[15] At the end of the conversion, these billets were backfilled with either contract employees or government civilians.

---

[13] The division-based structure was optimized to generate combat power for large-scale conventional operations. A division's structure achieved economies of scale with regard to key enablers—such as intelligence, surveillance, and reconnaissance; military police; fires; and logistics—which could be concentrated in support of committed maneuver brigades. In the distributed battlespace that characterized operations in Iraq, Afghanistan, and anticipated operational environments, in which almost all maneuver elements were continuously committed, achieving such economies of scale was unlikely. Ergo, the Army needed to create more enablers to support committed maneuver elements. The Army has described its modular transformation and associated initiatives in its annual posture statement since 2004 (see Brownlee and Schoomaker, 2004). For more on the modular transformation and supporting measures, see Donnelly, 2007, and Brown, 2011. For an assessment, see Johnson et al., 2012.

[14] See "Military to Civilian Conversions," an information paper in Harvey and Schoomaker, 2006, Appendix J. See also a similar document in Harvey and Schoomaker, 2007, Addendum Q.

[15] Wright and Reese, 2008, pp. 456–458, provides examples of how the Army relied on manpower from the generating force to fill advisor billets in Multinational Security Transition Command–Iraq early in the war.

RC forces provided an unprecedented—at least since the Korean War—share of the Army's deployed forces during this period. The USAR and the ARNG mobilized an average of 122,652 soldiers in FY 2004 and 119,433 in FY 2005, although the average declined significantly in FY 2006, to 88,058 (see FM&C, 2005a, p. 14; FM&C, 2006a, p. 14; FM&C, 2007a, p. 15).[16] RC units provided the full range of combat and combat service support capabilities in counterinsurgency operations in Iraq and Afghanistan, and several ARNG maneuver brigades conducted counterinsurgency operations in Iraq.

Figure 2.2 details the use of involuntary mobilizations of the USAR and ARNG to meet operational requirements over the 2001–2009 period, in which dramatic increases took place.

A reserve unit deployment included a year of deployment and additional time for training and other activities to prepare for deployment. Preparatory time ranged from about three to six months, with maneuver brigades requiring the most time to prepare for conducting counterinsurgency operations (Klimas et al., 2014, p. 5). This level of mobilization brought the RC's MOB:Dwell ratio to 1:3 (Defense Science Board, 2007,

Figure 2.2
Army RC Members on Active Duty from September 2001 to June 2009 in Support of Operations Noble Eagle, Iraqi Freedom, and Enduring Freedom

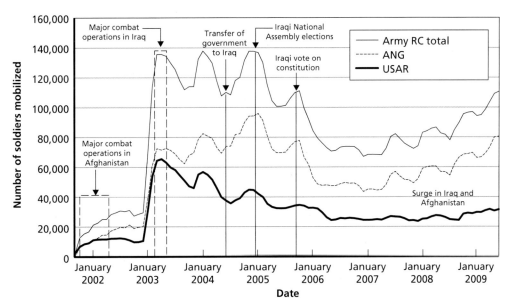

SOURCE: GAO, 2009, p. 22.
RAND RR1637-2.2

---

[16] FY 2004 mobilization figures are taken by subtracting the average of reported monthly strengths—which do not include mobilized reservists—from the annual average total, which includes mobilized reservists. In subsequent years, these tables list mobilized man-years.

p. 23). Note that the 1:3 ratio also was the objective for Regular Army forces, should demand ever decline to manageable levels (Harvey and Schoomaker, 2006, p. 10).[17]

The high level of RC use stressed the USAR and ARNG and appeared to approach the limits of congressional tolerance. Part of the strain was imposed by statute, or at least by its interpretation. In 2005, LTG James Helmly, chief of USAR, testified before the Military Personnel Subcommittee of the House Armed Services Committee that the RC was running out of soldiers it could deploy under the current partial mobilization authority. Helmly and other witnesses noted that the existing partial mobilization authority limited soldiers' cumulative mobilization time—and, by extension, that of units—to 24 months. This interpretation constrained the supply of RC forces even more than might be immediately apparent. A soldier who had been mobilized for a year's deployment plus a period of predeployment training could not then be remobilized for another 12-month deployment, because he would exceed the 24 cumulative month limit well before it was done. GEN Richard Cody, the Army Vice Chief of Staff, noted that the 1953 statute on partial mobilization authority was somewhat vague with respect to total mobilization time. He asked rhetorically, "it doesn't say when do you reset the clock. Is it two years, three years, four years or five years?" (U.S. House of Representatives, 2005, pp. 13–15).

While Army leaders—including the leaders of the RC—were primarily concerned about statutory limitations on mobilization, Congress—along with some in the RC—was concerned about the stress on RC soldiers and leaders and the potential breach of the implicit contract between the nation and its reservists. At the aforementioned hearing—entitled "The Adequacy of Army Forces"—Representative John Kline of Minnesota related the following anecdote:

> Last year we had a little discussion where the adjutant General from Minnesota, General Shellito, came out and talked to some of us and he said very bluntly that his soldiers, his national guard [sic] soldiers, did not enlist in the active Army, that they enlisted in the national guard and they are proud to serve there. They are proud to be called up and serve, but they didn't enlist in the active Army and they cannot be called upon to be continually called up. And it appeared to me that we were hearing a message that said the strain is getting very heavy on the members of the guard, their families and their employers. (U.S. House of Representatives, 2005, p. 21).

LTG Steven Blum, chief of the National Guard Bureau, had previously acknowledged that "the national guard [sic] is not in crisis, but it is significantly stretched" (U.S. House of Representatives, 2005, p. 12). For the most part, Blum and other high-level RC leaders were concerned about obtaining more resources and greater flexibil-

---

[17] It should be remembered, however, that the idea of rotational commitment of Army forces was just beginning to take hold at this time.

ity with respect to mobilization authority. Lower echelons may have been more concerned about the stress on RC soldiers. Certainly, their representatives were. In a Senate Armed Services Committee hearing later that year, Senator Carl Levin expressed the same concerns, albeit somewhat more forcefully:

> The only way that we have been able to meet our troop requirements in Iraq and Afghanistan is by mobilizing the overextended National Guard and Reserves. This has been done at a great cost to them, their families, and our communities. Governors are concerned about whether they will have National Guard personnel and equipment to respond to natural disasters. We continue to hear from employers about the adverse impact on small businesses and self-employed National Guard and Reserve members. Finally, some are wondering if the National Guard and Reserves will be ready the next time they are needed. In a memorandum to the Army Chief of Staff, the Chief of the Army Reserves said that "The Army Reserve is additionally in grave danger of being unable to meet their other operational requirements, including those in named operational plans and continental United States (CONUS) emergencies, and is rapidly degenerating into a broken force."

> The Chief of the National Guard Bureau recently stated that "My concern is that the National Guard will not be a ready force the next time it is needed, whether here at home or abroad." Our overreliance on the Guard and Reserve may have severely impacted them as effective military units. (U.S. Senate, 2005, p. 4)

That level of concern led to the 2005 establishment of the Commission on the National Guard and the Reserve and the initiation of a Defense Science Board study, *Deployment of Members of the National Guard and Reserve in the Global War on Terrorism* (Defense Science Board, 2007; Commission on the National Guard and the Reserves, 2008).

Yet for all this concern, the RC did not actually "break" in any meaningful sense of the word, and Congress did not withhold support for their further mobilization and deployment. In a hearing on "Adequacy of Army Forces," General Blum affirmed that, if properly resourced, the ARNG could sustain having 25 percent of its force committed indefinitely, a percentage equivalent to the 1:3 MOB:Dwell ratio at which they were operating (U.S. House of Representatives, 2005, p. 22). Operating at this tempo for about three years—the period of highest use between the end of 2003 and the end of 2006—approached, but did not reach, the limits of the RC's capacity to sustain operations or congressional willingness to support its continued mobilization.

Increasing Army Size: Headwinds in Recruiting and Retention Efforts

The other major Army initiative was increasing the size of the Regular Army. The conditions under which it tried to accomplish that increase, however, were substantially more challenging than those that had obtained in the previous period. The wars in Iraq and, to a lesser extent, Afghanistan started going poorly. Casualties mounted; deaths in Iraq and Afghanistan climbed to a total of 2,575 by the end of 2006 (DMDC, 2015a). As the war deteriorated, public support eroded. By December 2006, 53 percent of respondents to a Gallup poll concluded that it had been a mistake to send troops to Iraq in the first place, a substantial increase from the 42 percent who thought so in January 2004 and the 23 percent who had disapproved of the invasion in March 2003 ("Iraq," undated). At the same time, the economy continued to improve. Unemployment declined from its 2003 average of 6.0 percent to 4.6 percent by the end of 2006 (BLS, 2014). Concurrent with these factors, propensity to enlist declined, from 15 percent for all young people in May 2004 to 10 percent in June 2006 (Commission on the National Guard and the Reserves, 2008). The Army faced substantial headwinds just in terms of maintaining the force that it had, let alone increasing its size.

Under these conditions, the Regular Army struggled to maintain its end strength at around 500,000 soldiers. End strength in FY 2003 had risen rapidly to 499,000, 19,000 soldiers more than the authorized end strength of 480,000 (FM&C, 2004, p. 5). In authorizing the January 2004 temporary end strength increase of 30,000, Congress required it to be achieved by FY 2006. By that time, however, end strength had increased to only 505,402, an increase of just over 6,000 over three years (FM&C, 2007a, p. 6). To some degree, the modest magnitude of this increase may have owed something to the Army's reluctance to alter its long-term program for temporary expansion. The 2004 National Defense Authorization Act, which had authorized the increase in end strength, had initially stipulated that the Army pay for any end strength over and above the Army's permanent end strength authorization of 482,400, at least in part, by reallocating funds from other purposes (CBO, 2006).

The Army also encountered difficulties in recruiting and retention efforts. It struggled to meet its accession goals, missing its goal for FY 2005 by more than 6,000 soldiers. It also fell somewhat short of its quality goals: The DoD goal was—and remains—that 90 percent of recruits have high school diplomas, but in 2005, only 87 percent of recruits had this credential (FM&C, 2006a, p. 7). The Army was able to achieve a numeric goal of 80,000 accessions in FY 2006, but only at the expense of reducing quality still further. Only 81 percent of accessions had high school diplomas; only 61 percent scored in Categories I–IIIA of the AFQT (FM&C, 2007a, pp. 6–7). The Army also relaxed standards with regard to age, conduct, and physical condition, granting waivers to nearly 20 percent of new recruits (Kapp and Henning, 2008; U.S. House of Representatives, 2007b, p. 8).

Another indicator of the Army's difficulty in recruiting was the size of its delayed entry pool (DEP). The DEP consists of soldiers who sign a contract in one year to enlist the next; this is useful for maintaining the flow of recruits into the training base. The Army normally prefers to maintain a DEP of approximately 35 percent of its accession goal. At the beginning of FY 2005, however, the DEP fell to 18 percent of the year's accession goal. In FY 2006, it fell further, to 12 percent (Kapp, 2006; Kapp and Henning, 2008). We should note, however, that this drop alone is not indicative of Army recruiting shortfalls; the Army instituted a "quick ship" bonus program providing incentives to recruits to enter service immediately. This bonus program may have increased the number of recruits entering immediately at the expense of meeting DEP goals (Kapp, 2006). Retention also initially suffered but recovered rapidly. In 2004, the Army fell short of retention goals for initial term, midcareer, and career soldiers. In 2005, the Army achieved or exceeded these three goals. By FY 2006, it did so substantially, exceeding retention goals by almost 14 percent (OUSD[C], 2005; OUSD[C], 2006).

The Army's recruiting and retention difficulties were not confined to the Regular Army; both the ARNG and USAR struggled to maintain end strength, accession, and retention targets. The ARNG finished FY 2004 with almost 8,000 fewer soldiers than its authorized end strength of 350,000. The shortage increased to 17,000 in 2005; then, end strength climbed to a little more than 346,000 (FM&C, 2005c, p. 5; FM&C, 2006c, p. 5; FM&C, 2007c, p. 6). USAR strength declined rapidly from 204,000 in FY 2004 to about 189,000 in FY 2005, then remained at that level through the remainder of the period (FM&C, 2005b, pp. 10–11; FM&C, 2006b, p. 9; FM&C, 2007b, p. 10).

Recruiting and retention conditions had become unfavorable, but decisions to reduce the Army's recruiting force in 2003 and 2004 exacerbated problems in 2005. Since FY 2002, the Army had steadily decreased the number of recruiters, from a peak of 6,367 to a relative nadir of 5,109 in FY 2004 (CBO, 2006, p. 9). However, the Army needed the midgrade noncommissioned officers (NCOs) in its recruiting force to accomplish its plan to increase the size of operating forces by drawing manpower from the Army's generating force. According to a recruiting command official, the plan had been even more drastic. Planners had assumed that the conditions underpinning the Army's success in recruiting in FYs 2002 and 2003 would continue and had planned to reduce the number of recruiters to about 4,900. The February 2004 decision to increase the accession goal to 80,000 gave manpower planners relatively little time to prepare for the increased mission. Army officials had to identify the additional recruiters, divert them from their intended assignments, and get them trained and to the field with very little notice. The Army restored its recruiting force to 5,953 in FY 2005. The recruiting force for the RC followed similar patterns. In all cases, it proved difficult to recover from opportunities lost in FY 2004 and reflected in the reduced DEP.[18]

To restore accessions and retention to the levels required to sustain the force at the desired end strength, the Army substantially increased recruiting and retention resources. In addition to the increased recruiting force, the Army expanded both the eligibility for enlistment and retention bonuses and the amounts of those bonuses. For example, the maximum enlistment bonus doubled in FY 2006 for selected occupa-

---

[18] U.S. Army Recruiting Command official, telephone interview with M. Wade Markel, May 5, 2015, and CBO, 2006, p. 9. See also Note 8 in Kapp and Henning, 2008.

tions (CBO, 2006, p. 7). Total Regular Army enlistment and reenlistment bonuses (in terms of constant FY 2015 dollars) increased from $348 million in FY 2003 to more than $1.3 billion in FY 2006. As shown earlier, in Figure 2.1, the number of contracts signed with non–prior-service recruits surged between FYs 2003 and 2006. The share of these contracts that included an incentive went from 49 percent in FY 2003 to 67 percent in FY 2006, and the average incentive value more than doubled. RC selective reserve retention incentives increased from $216 million to $984 million over that same period.[19] Expanded eligibility criteria made more soldiers eligible for bonuses.

January 2007–December 2010: Duration of the Grow the Army Initiative

The Iraq surge began in 2007. By year's end, Army forces there had increased from 119,500 to 138,500. Army forces committed to Afghanistan increased by almost 4,000 soldiers as well, to 19,200. All told, the Army had about 158,000 soldiers deployed to both conflicts by year's end. Throughout the remainder of the period, Army strength in Iraq declined, while the number of soldiers deployed to Afghanistan increased. By September 2010, deployed Army strength had fallen to about 123,000 soldiers, still higher than it had been in 2004 (DMDC, undated). Maintaining that level of strength in the field required the Regular Army to maintain a BOG:Dwell ratio of about 1:1.3, with the RC remaining slightly below 1:3 (U.S. Senate, 2009, p. 7).

Just as important, Army and defense leaders had come to believe this level of commitment represented a "new normal" in what Acting Secretary of the Army Pete Geren called "an era of persistent conflict" at his confirmation hearing in June 2007 (U.S. Senate, 2007, p. 710). The 2008 edition of Field Manual 3-0, *Operations*, defined the period as one "of protracted confrontation among states, nonstate, and individual actors increasingly willing to use violence to achieve their political and ideological ends" (Field Manual 3-0, 2008, "Foreword"). In early 2007, the Army Chief of Staff, General Schoomaker, envisioned providing between 18 and 19 BCTs annually, along with other capabilities. By 2011, Army Regulation (AR) 525-29, 2011, p. 3, stated that the Army had "1 corps headquarters (HQ), 5 division HQs, 20 BCTs, and approximately 90K [thousand] enablers," a total force of about 170,000. Actual demand never reached these levels, but they reflect the shift in contemporary thinking of key Army decisionmakers.

The measures on which the Army had embarked to meet this demand in 2004–2006 largely reached fruition in this period. In 2010, the Army reported that it had

---

[19] For FY 2003 figures, see FM&C, 2005a, pp. 85, 89; FM&C, 2005c, p. 83; FM&C, 2005b, p. 105. For FY 2006 figures, see FM&C, 2007a, pp. 86, 89; FM&C, 2007b, p. 110; FM&C, 2007c, p. 86. Dollar estimates were converted into constant FY 2015 dollars using the operation and maintenance deflators from OUSD(C), 2014, Table 5-6.

accomplished 88 percent of its modular transformation (McHugh and Casey, 2010, p. 11). Efforts to transfer military billets from the generating force to the operational Army waned in this period but did not reverse. The Army continued to rely heavily on contractors to perform ancillary operational functions, such as logistics and site security, and substantially increased spending on service contracts to replace military manpower in the generating force. It continued to draw routinely on the RC, leading to recognition of an "operational" RC in DoD policy. But the centerpiece of the Army's effort to meet current demand and enable the Army to do so without undue strain was the Grow the Army initiative. As in the previous section, we do not address modular transformation so that we can focus on the measures taken to increase Army capacity, particularly the Grow the Army initiative.

Revisiting Transformation of the Generating Force

At this time, efforts to transfer spaces from the generating force to the operational Army appeared to reach the limit of the possible. Military-to-civilian conversions waned as a growth option, although possible conversions may have reached their outer limits. The 2008 *Army Posture Statement* was the last to mention military-to-civilian conversions, noting that about 10,000 such conversions had been made (Geren and Casey, 2008, p. 13). In 2010, GEN Martin Dempsey, then the commander of U.S. Army Training and Doctrine Command (TRADOC), wrote a letter to the Army Chief of Staff indicating that manning shortfalls put TRADOC's ability to execute its core functions at risk. TRADOC was one of the larger generating force commands and the one that depended most heavily on military manpower. TRADOC had prioritized training over its other functions in the allocation of military manpower, accepting risk with regard to doctrine and force development functions and relying on contractors to perform training when and where possible (GAO, 2011, p. 1). By February 2010, Dempsey had concluded that this temporary expedient was putting TRADOC's capabilities at risk over the long term. The only other large sources of military manpower were the U.S. Army Medical Command and, to a lesser extent, the Army Forces Command, where soldiers played the critical role of ensuring the readiness of deploying forces.

The Army continued to rely on the RC to provide the forces required overseas but used fewer reservists. In January of 2007, newly appointed Secretary of Defense Robert Gates promulgated a policy expanding access to the RC. Gates' memorandum, "Utilization of the Total Force," interpreted the partial mobilization authority's 24-month limit on mobilization as applying to consecutive months, rather than cumulative months.[20] Under the new authority, units could in theory be mobilized for 24 months, demobilized for a day, and then remobilized. The new policy tried to soften its effect by limiting mobilization periods to one year. At the same time, RC mobilization declined further, from 88,058 in 2006 to 69,980 in 2007. It briefly rose again, reach-

---

[20] See Appendix E, "Utilization of the Total Force," in Defense Science Board, 2007.

ing 88,454, before resuming its uninterrupted decline.[21] During this period, RC forces were employed principally in combat support and combat service support roles. For example, while four ARNG BCTs deployed to Iraq in FY 2009, they were employed as security forces (FM&C, 2009, p. 3). Even at these lower levels, Army leaders still estimated that RC MOB:Dwell continued to hover at around 1:3, which GEN Peter Chiarelli confirmed in April 2009 (U.S. Senate, 2009, p. 6).

Over this time, the RC also began to be seen as an essential piece of a Total Army operational force rather than as part-time "backup" soldiers. In 2007, key officials and decisionmakers had expressed doubt as to how long the RC could maintain this pace (Commission on the National Guard and Reserves, 2008, pp. 179–180). Within two years, the RC had proven to outside stakeholders that it was capable of functioning as an "operational reserve." In 2008, this evolution was formalized in a DoD directive on "Managing the Reserve Components as an Operational Force" that envisioned continuous employment of the RC to supplement and complement Regular Army capabilities and capacity (DoD Directive 1200.17, 2008). Elected officials went from voicing anxiety over operational stress to celebrating the RC's new responsibility to "support, augment, and assist our active duty forces on a routine and continuing basis," in the words of Representative Solomon Ortiz, Chairman of the House Armed Services Committee's Subcommittee on Readiness (U.S. House of Representatives, 2010, p. 1).

Growing the Army in Changing Circumstances

Secretary of Defense Gates' other key decision was to expand, or Grow the Army. In January 2007, Gates authorized an increase of 65,000 in permanent authorized AC Army end strength over five years, to a total of approximately 547,000. Initially, the actual increase was somewhat smaller and left the Army with a need to grow by approximately 40,000. The plan at the time was to grow by approximately 7,000 per year, achieving the increase by the end of FY 2012 (FM&C, 2009, p. 4). To cope with continued shortfalls in deploying units, Secretary of Defense Gates and Chief, Joint Chiefs of Staff ADM Michael Mullen announced a second Grow the Army Campaign in July 2009. This campaign would provide a temporary increase of 22,000 in AC end strength in addition to the goal of an end strength of 547,000. The increase would be phased in over three years before its eventual expiration in 2013 (DoD, 2009, p. 8).

Initially, conditions for achieving the increase in the size of the Regular Army were somewhat daunting. In December 2006, the Iraq Study Group released its report, contributing to a growing sense that the war was not going well and that a substantial change in strategy was needed. Its opening sentence stated "The situation in Iraq is grave and deteriorating" (Hamilton and Baker, 2006, p. viii). Public support for the war dropped substantially after the report's publication. A bipartisan coalition of legis-

---

[21] See the "actual" reported figures in FM&C, 2008a, p. 14; FM&C, 2009a, p. 12; FM&C, 2010a, p. 12; FM&C, 2011a, p. 18.

lators promoted the Iraq Study Group Recommendations Implementation Act of 2007 in an attempt to force a change in U.S. strategy and reduce the U.S. commitment to the war. The bill ultimately did not become law but may well have had enough support to pass both houses of Congress (Tama, 2007). In a January 2007 Gallup poll, 58 percent of Americans surveyed believed the whole effort had been a mistake ("Iraq," undated). Casualties in Iraq and Afghanistan continued to mount, with U.S. forces suffering another 1,020 deaths over the course of 2006 (DMDC, 2015a; DMDC, 2015b). In addition, unemployment stood at 4.6 percent (BLS, 2014). By June 2007, propensity to serve had declined to 9 percent (Commission on the National Guard and Reserves, 2008, p. 77).

In 2008, the financial crisis immeasurably improved external recruiting and retention conditions. Unemployment increased, initially to 5.8 percent in 2008, then to an annual average of 9.6 percent in 2010. Meanwhile, the Iraq war seemed to have taken a turn for the better. By February 2008, 40 percent of respondents to a Gallup poll believed the "Surge" had improved conditions and implicitly improved U.S. chances of success ("Iraq," undated). In 2008, OIF casualties were one-third of what they had been in 2007, the deadliest single year of that war. By June 2010, male propensity for military service had rebounded from its December 2007 nadir to just above 15 percent, although it would never reach its post-9/11 high again (see Carvalho et al., 2010, Fig. 3-5, pp. 3–14).

Internally, the Army also increased the resources available to recruit and retain soldiers. Between FY 2007 and FY 2009, the number of recruiters increased by more than 1,000.[22] In terms of constant FY 2015 dollars, the Army spent between $1.2 billion and $1.3 billion on recruiting and retention incentives each year from FY 2007 to FY 2009. The number of contracts signed with non–prior-service recruits that included an incentive continued to rise during this period, with 79 percent of these contracts including an incentive in FY 2009 (Figure 2.1). The Army also increased the number of military occupational specialties (MOSs) for which it offered reenlistment bonuses from 12 to 76, encompassing virtually the Army's entire enlisted force (FM&C, 2007a, pp. 83–89; FM&C, 2009, pp. 73–80; FM&C, 2010a, pp. 72–78; FM&C, 2011a, pp. 68–73; FM&C, 2012, pp. 86–91).

Favorable conditions and strenuous effort enabled the Army to accelerate its timetable for growing the force. Originally, the Army had intended to reach its target of 547,000 by the end of FY 2012. In this more favorable environment, the service exceeded that goal by September 2009, reaching an end strength of 552,465 (FM&C, 2011a, p. 7). In 2009, the Army also began to meet its quality goals again, with more than 95 percent of accessions having a high school diploma—compared with a goal of 90 percent—and 66 percent scoring in Categories I–IIIA on the AFQT, compared with a goal of 60 percent. The size of the DEP also climbed to 33,000 by FY 2010's

[22] Joint Advertising, Market Research & Studies data, provided to authors by Army Marketing Research Group.

end. The large DEP indicates that the Army might have been able to recruit still more had it proved necessary, especially if the quality mix had remained as it had been in the 2005–2007 time frame. During this period, the Army also exceeded retention goals in every year (see Kapp and Henning, 2009, Table 5, p. 10). The Army's RC experienced similar degrees of success (see Kapp and Henning, 2009, Tables 2–4, pp. CRS-4 to CRS-7). In this new environment, the challenge shifted from finding recruits to accommodating them.

However, the effect on operational capacity may have been less than meets the eye. Even though the Army attained an end strength of 552,000 soldiers by the end of FY 2009, about 30,000 of them were essentially unavailable to fill operating force units. In April 2009, the Vice Chief of Staff of the Army noted that the Army had

- 9,500 wounded warriors assigned to warrior transition units
- 2,300 soldiers working as cadre or health care providers
- 10,000 nondeployable for health, legal, or other reasons
- 10,000 individual augmentees (U.S. Senate, 2009, p. 6).

Obviously, the 10,000 individual augmentees were contributing to the war effort. The salient point is that Army force structure, from which end strength requirements were derived, did not anticipate or include the need for these augmentees. The problem may have been even worse than originally reported. A 2011 U.S. Army War College research paper found that nondeployable rates had been increasing steadily since at least 2007, when 9.9 percent of the soldiers in BCTs had been nondeployable. By 2011, the proportion of nondeployable soldiers in BCTs had reached 14.5 percent (Arnold et al., 2011, p. 3).[23]

Since then, the Army has undertaken aggressive efforts to increase medical readiness and decrease the number of nondeployable soldiers in the ranks. The U.S. Army Medical Command initiated an integrated campaign plan to identify medical readiness problems, treat soldiers who could be restored to duty, and transition those who could not be restored to deployable status from the service (U.S. Army Medical Command, 2011; Schoomaker et al., 2011). Concurrently, the tempo of deployments declined even more steeply. The proportion of nondeployable soldiers remained high but had declined significantly from its peak (Cox, 2015).

---

[23] Clearly, these numbers are difficult to reconcile. The Army War College study implicitly indicates that more than 10,000 soldiers would have been nondeployable in 2009, since there would have been around 150,000 Regular Army soldiers in BCTs (43 BCTs with approximately 3,300 soldiers each) at that time, for an estimate of about 15,000 soldiers (Arnold et al., 2011). Moreover, non-BCT units were also likely facing challenges with nondeployability. The main point here is that the Army faced a significant challenge with nondeployable soldiers throughout the period.

From January 2011 Forward: Drawing Down

During this period, the Army found its focus shifting abruptly from maintaining the force it had built to downsizing. Waning operations in Iraq and the sharp time limitations on the Afghan surge reduced current operational demands for Army forces. Meanwhile, the country's fiscal circumstances and sequestration compelled reducing the cost of America's armed forces, principally the Army. Because personnel costs are a large portion of defense spending, force size needed to come down as well. DoD and administration priorities also had shifted away from the types of operations fueling Army growth from 2004 through 2010. On January 26, 2012, the new DSG stated that U.S. forces would no longer be sized to conduct large-scale, prolonged stability operations (DoD, 2012; Shanker and Bumiller, 2012). On January 27, 2012, the Chief of Staff of the Army, GEN Ray Odierno, announced that AC Army end strength would be reduced by as many as 80,000 personnel by the end of 2017 (Odierno, 2012; Lopez, 2012; Banco, 2013). In July 2013, Secretary of Defense Chuck Hagel suggested that Army end strength might be reduced to between 420,000 and 450,000 in the AC and to between 490,000 and 530,000 in the RC (DoD, 2013). These kinds of reductions were reiterated in the administration's FY 2015 Budget Guidance (Feickert, 2014). The problem was no longer growing the Army, but reducing it prudently.

During this time, DoD and the Army did not support drastic force reductions below approximately 450,000 active Army end strength. They also expressed concerns about sequestration-level cuts requiring a further reduction to 420,000. The Army stated that the 450,000 active Army; 335,000 ARNG; and 195,000 USAR end strengths constitute the "smallest acceptable force to implement the defense strategy" (Feickert, 2014, p. 12, citing Sprenger, 2014) and described a 420,000-soldier force as providing "insufficient capacity" that "cannot implement [the] defense strategy" (U.S. Army, 2014, p. 5, cited in Feickert, 2014).

Beyond Soldiers: Using Contractors to Augment Operational Capacity

Although much of this chapter has focused on Army forces, there is more to operational capacity than soldiers and the military formations of which they are a part. From the outset, the Army augmented its operational capacity in Iraq and Afghanistan with contractors. Contractors performed a variety of functions, ranging from basic sustainment, such as operating dining facilities, to providing security for military facilities and U.S. government personnel from other agencies. Many in the last category, known as private security contractors, did not actually work under Army contracts but did perform functions that very well could have fallen to Army personnel had the agency been unable to contract for their protection. For the first half of OIF, data on the number of contract personnel were hard to find. In a letter to Representative Henry

Waxman, Chairman of the House Government Reform and Oversight Committee, the Deputy Assistant Secretary of the Army for Policy and Procurement reported that the Army had, by itself, 96,130 contractors supporting its operations in Iraq (Ballard, 2007). In September 2007, the U.S. Central Command began reporting the total number of contractors supporting DoD. Table 2.1 depicts the annual figures. In addition to the number of contractors, the table also reports the total number of military personnel deployed. While such calculations are inherently problematic, it would have taken between 235,000 and 508,000 military service members to produce the same level of operational capacity at a BOG:Dwell rate of 1:1.

Contract support thus played an indispensable role in providing operational capacity. However, such support came at a cost. Some of the costs are intuitive, inherent in the perennial triad of waste, fraud, and abuse. Estimates of waste and fraud in Iraq and Afghanistan contracts range as high as $60 billion, in nominal dollars. Contractors also have different fiduciary responsibilities; their legal responsibility is to produce maximum profit for their shareholders or owners at minimum risk, not necessarily to take whatever steps are necessary to accomplish the mission. Sometimes the specifications of their contracts may actually conflict with mission requirements at the time (Commission on Wartime Contracting in Iraq and Afghanistan, 2011, p. 1). For the most part, assessments attributed many of the problems with contract support to inadequate military capacity for management and oversight for contracting in a contingency environment (Schwartz and Church, 2013, p. 8; Commission on Wartime Contracting in Iraq and Afghanistan, 2011, p. 1).

Table 2.1
Department of Defense Contractors in Afghanistan and Iraq, 2007–2013

Report Date	Afghanistan	Iraq	Total Contract Support Personnel	Total Military Personnel Deployed	Contractors as a Percentage of the Total Force
September 2007	29,473	154,825	184,298	243,740	43
September 2008	68,252	163,446	231,698	222,700	51
September 2009	104,101	113,731	217,832	230500	49
September 2010	70,599	74,106	144,705	202,100	42
September 2011	101,789	52,637	154,426	201,400	43
September 2012	109,654	10,967	120,621	146,712	45
March 2013	107,796		107,796	132,048	45

SOURCES: Numbers of contract support personnel in Afghanistan and Iraq are from Schwartz and Church, 2013, pp. 24–25. Numbers of deployed military personnel are from DMDC, undated.

Summary and Conclusion

This chapter has derived insights from the Army's recent historical experience about its future potential to expand rapidly when called on to do so. That experience consists of four major periods:

- 9/11 through December 2003, when popular support and favorable conditions enabled the Army to meet relatively modest demands, limited in anticipated duration if not necessarily scale
- January 2004 through December 2006, when the Army struggled to anticipate and meet the ambiguous demands of an unpopular and worsening war
- January 2007 through December 2010, as the Army expanded its capacity, this time under favorable conditions, to provide the ability to sustain large-scale combat operations indefinitely if need be
- January 2011 onward, as demand declined, strategy changed, and the Army began downsizing.

Our key focus as we examined these periods was to identify any limits on the Army's ability to expand when called on to do so. That focus informs the following findings.

Army Capacity Depended Heavily on Conditions

This finding emerges most clearly from the efforts to increase Regular Army end strength but extends beyond that context. Civilian and military assumptions about the size and scale of the conflict affect force planning. During the 15 years we examined, policymakers shifted from assuming modest force requirements for a smaller-scale conflict for a short time to the sizable troop levels required to conduct two simultaneous and protracted conflicts. Also, policymakers assumed that existing authorities for expanding Army manpower (e.g., mobilization authorities, end strength levels, stop loss) would be sufficient to meet post-9/11 operational demands. Although there was criticism of some policy options used, members of Congress and the public generally granted policymakers wide latitude in selecting the authorities best suited to meet operational demands. In future conflicts, the Army may face either a permissive or a restrictive environment for exercising available policy options.

External conditions also affect the ability to Grow the Army. When the Iraq and Afghanistan wars were popular and seen as successful—even if only relative to their previous trajectory—maintaining end strength and quality posed no extraordinary challenges. When the wars became unpopular and were seen to have been a mistake—a circumstance reflected in but probably not entirely the result of high casualty rates—the Army struggled to provide capacity. Parents, teachers, and other "influencers" withdrew their support for military service, and congressional scrutiny heightened, especially with regard to the use of the RC. The state of the economy also played an important role: When unemployment is low, as it was in the 2004–2006 time frame, recruiting and retention are hard. When unemployment is high, as it was after 2008, recruiting and retention become much easier.

Conditions govern more than recruiting and retention. Had the war not commenced with the terrorist attacks of 9/11 and had the Bush administration not received an electoral mandate in favor of the use of force against Iraq, it might not have been so easy to draw on the RC for so long. The Secretary of Defense reinterpreted his mobilization authority without consulting Congress. Had the wars begun otherwise, that option may not have existed. The Army and DoD also depended heavily on contractors to provide operational capacity, with the number of contractors often equaling or exceeding the number of soldiers committed. Had the enemy been more capable, perhaps operating in more favorable terrain, it might not have been possible to rely on contract support to that extent, requiring a greater commitment of military personnel.

Many of these conditions are, however, largely outside the Army's immediate control. Some of these conditions may be inevitable, such as difficulty adapting to insurgencies. Military organizations inevitably struggle to recognize the existence of insurgencies and adapt to them. The state of the economy, which seems to be the most important determinant of the Army's ability to expand, falls completely outside the Army's control. The Army, as an organization, might not be able to directly influence public opinion on a particular war or major operation, but public support may not be a completely exogenous variable. Several studies have indicated that presidential leadership can influence public perceptions of the war to a limited degree (see, for example, Eshbaugh-Soha and Linebarger, 2014, and Tedin, Rottinghaus, and Rodgers, 2011).[24] Although the Army cannot control the circumstances that govern its ability to expand, it can at least plan realistically for the conditions it may encounter.

The Regular Army Was Able to Sustain a BOG:Dwell Ratio Approaching 1:1 More or Less Indefinitely

From the invasion of Iraq through the end of the Afghan surge, the Regular Army operated more or less at a 1:1 BOG:Dwell ratio. That is not to say that there were no problems. Suicide rates increased, as did nondeployability rates and misconduct. Moreover, the Army was forced to obligate vast sums of money to provide incentives for recruiting and retention. For all that, the Army experienced no dramatic reduction in effectiveness or efficiency during that time that significantly degraded operational capability or capacity.

---

[24] These conclusions are tentative, however. Eshbaugh-Sona and Linebarger, 2014, indicated that presidential leadership—measured in terms of rhetorical tone—mostly affects public perception of the president. Similarly, Tedin, Rottinghaus, and Rodgers, 2011, notes that public *policy* preferences—what should be done about the war in question—are the *least* susceptible to presidential persuasion.

A 1:3 MOB:Dwell Ratio Probably Represented the Limit of the Army RC Capacity for Operational Support in Extended Operations

In 2004 and 2005, the Army was mobilizing close to 120,000 RC soldiers annually, a MOB:Dwell ratio reported as 1:3. In this context, Congress repeatedly expressed great concern about the tempo at which the Guard and Reserve were being used; Congress even considered a law that would have required the Army to provide a certain amount of dwell time, especially for reservists. Even though Congress did not pass that law, the average number of soldiers mobilized had dropped by about 50 percent by 2006, and the levels were even lower in 2007. Subsequently, mobilizations did not increase much beyond 90,000. Even though the Army continued to assess the MOB:Dwell rate at 1:3, reduced demands clearly lowered stress on the RC. There is probably some leeway there; Congress did not forbid DoD to employ rates above 1:3, and the "correct" MOB:Dwell ratio is still debated. Still, the Army was no longer attracting unfavorable attention once demand declined somewhat.

The Regular Army May Have Approached the Limit on Its Ability to Expand in Unfavorable Conditions Between FY 2004 and FY 2006

During this period, unemployment was low; the wars were unpopular; and the Army struggled to provide required capacity. As we noted, it was unable to source a request for forces for Afghanistan in 2005. Congress was scrutinizing the employment of reserve forces, and the Army struggled to expand by 6,000 soldiers over three FYs; Army end strength had reached 499,000 by the end of FY 2003. It fell short of its accession target in FY 2005 by almost 7,000 soldiers, and retention fell below targets initially. The Army was able to meet recruiting and retention targets by lavishing enlistment and reenlistment bonuses on recruits and soldiers and by reducing quality standards for accessions. The small size of the DEP might indicate that the Army had little headroom for increasing size under good economic and poor strategic conditions. To the extent that the foregoing description forms a reasonable set of assumptions for the future, the Army may only be able to recruit up to 80,000 soldiers annually under such conditions, with the recruits representing a lower-quality mix.

Nevertheless, some indications suggest that this apparent limit may not be absolute. First, accessions well above 100,000 were common before the previous drawdown, in the 1980s. Second, contemporary observers believed that the Army had incurred some of the risk through its own mistakes, reducing its recruiting force just when it became needed. Even the small DEP is attributable to the structure of enlistment incentives. "Quick ship" bonuses rewarded immediate enlistment, but there was no comparable incentive to join a waiting list. Third, it seems clear that Army and DoD officials were loath to expand the Army substantially. Had the Army fully exploited its authority to increase end strength, it would have had to find the funds elsewhere in its base budget, forcing direct conflict between short- and long-term priorities. Consequently, the Army might have been reluctant to divert resources from the other priorities to recruiting and retention. Had those activities received the same priority that they did in 2007 and afterward, it is not inconceivable that the Army would have been able to increase recruiting to a degree. Yet even though the Army may have been able to improve its performance, that performance probably represents a reasonable point of comparison for future estimates.

The Army Could Probably Expand More Than It Did Between FY 2007 and FY 2009 Under Favorable Conditions

Conversely, even though the Army accomplished its objective for end strength growth—to 547,000—earlier than initially planned, it might have been able to do even better had it wanted to. The Great Recession presented a uniquely favorable set of conditions, as unemployment skyrocketed. Including recruits entering the DEP, the Army was signing contracts with more than 100,000 recruits annually.[25] The economic picture also allowed the Army to raise quality standards, limiting intake. Had the Army not begun to restrict eligibility criteria, it might well have secured even more recruits.

The Army Had to Relax Eligibility Standards to Increase Capacity in Unfavorable Conditions

After failing to achieve its accession objective in FY 2005, the Army deployed substantial resources to recover from the shortfall. It deployed additional recruiters and increased enlistment and reenlistment bonuses by nearly 60 percent. Even so, it met its FY 2006 objective only by accepting a substantially reduced proportion of high school graduates and candidates who scored in Categories I–IIIA on the AFQT and by accepting an increased proportion of soldiers with moral and other waivers. Reduced levels of quality continued through FY 2008, until the Great Recession that began in 2008 had its full effect. It is possible, however, that quality levels could have remained high had the Army offered larger financial incentives. Looking forward, only the most favorable circumstances from a recruiting and retention perspective promise to allow the Army to increase accessions substantially and maintain quality simultaneously, unless much higher financial resources are allocated to these efforts.

Wars End; Demand Declines

When the Army expanded, it did so based on the assumption that the Army would have to maintain high levels of operational commitment indefinitely. That assumption undoubtedly informed the force mix and the kinds of incentives offered to recruits and reenlistees, among other things. Yet, as the wars declined, so did support for a larger Army, as it has after every war the United States has fought.

---

[25] Data provided to authors by Army Marketing Research Group.

CHAPTER THREE

Regeneration Scenarios

This chapter describes the regeneration scenarios we used in our analysis. It explains the rationale for their adoption and the underlying analysis. These scenarios are illustrative only. In reality, the actual speed and scale of regeneration required will depend on the nature of the contingency for which forces are being raised.

We focused on scenarios requiring rapid expansion to meet the demands of large-scale, protracted contingency operations of approximately the same scale as those in Iraq and Afghanistan since 9/11. We did so for three reasons. First, these are the kinds of contingency operations for which the 2012 DSG proposed regeneration—or "reversibility"—as a hedge (DoD, 2012). Second, using these scenarios enabled the research team to compare results against historical experience, making it possible to identify relevant variables and to reconsider assumptions. Third, in contrast to current planning scenarios used in DoD's support to strategic analysis, these examples are unclassified, allowing wider dissemination of the resulting analyses.

The research team developed two major scenarios, basing them on alternative force structure and end strength options articulated in Army policy documents. Both scenarios presuppose a large-scale conventional conflict followed by a large-scale, long-term counterinsurgency effort. Both scenarios presuppose a period of five years from the onset of conflict until the last unit is available for employment. A previous RAND study found that most insurgencies last about ten years; the Army should be able to generate the strength to prevail by about the halfway point (Connable and Libicki, 2010, p. xii). The scenarios differ with respect to their starting conditions. The first scenario is an expansion from an Army with an authorized Regular Army end strength of 450K and a total end strength across all three components of 980K, which was the Army's preferred target for the drawdown (McHugh and Odierno, 2015, p. 2). The second scenario involves an expansion from a Regular Army end strength of 420K and a total end strength of 920K, which were the Army's targets if sequestration remained in effect (McHugh and Odierno, 2014, p. 2).[1] Table 3.1 depicts these end strength options.

---

[1] The 2015 posture statement does not discuss this option.

Table 3.1
Army End Strength Options Considered

Component	980K Option	920K Option
Regular Army	450,000	420,000
Army National Guard	335,000	315,000
Army Reserve	195,000	185,000
Total	980,000	920,000

SOURCES: McHugh and Odierno, 2014; McHugh and Odierno, 2015.

Estimating Required Operational Capacity

In developing these scenarios, we postulated that the Regular Army would have to expand to a degree that enabled the Army as a whole to produce the same amount of capacity as it did at the end of FY 2009, when it reached its targets for expansion under the Grow the Army campaign.[2] Operational capacity is a function of available end strength; the number of soldiers in the transients, trainees, holdees, and students (TTHS) account who are unavailable for deployment; the number committed to the Army's generating force, also assumed to be unavailable for deployment; and the force management policies governing the ratio of time deployed (BOG) to time at home station (Dwell).

Our basic logic was to calculate the shortfall between the 980K and 920K options and the 2009 baseline capacity. Obviously, doing so will require restoring the Regular Army to at least its 2009 baseline. Still more Regular Army strength will be needed to replace the RC strength that will have been lost as well.[3]

We began by calculating the raw shortfall for the 980K and 920K options—the difference between the end strength in 2009 and the end strength in each option—for the Regular Army, USAR, and ARNG (columns 3 and 6 of Table 3.2). We then estimated the active duty capacity that would be required to replace the shortfall.

DoD policy was to strive for AC BOG:Dwell ratios of 1:2 during "surge rotation" periods and 1:3 at all other times. For the RC, the ratio of time mobilized to time not

---

[2]   As discussed in Chapter Two, contractors also played an important role in operational capacity during OEF and OIF. We do not explicitly model the regeneration of contractors here but do discuss the role of contractors in Chapter Five.

[3]   With the prescribed rotation rates—including overlap and postmobilization training—it takes more than six RC soldiers to maintain one deployed but only slightly more than two Regular Army soldiers to maintain one deployed. In this case, expanding the Regular Army is thus the most efficient way to increase Total Army capacity.

Table 3.2
Estimating Additional Regular Army Strength Needed to Generate Capacity Equivalent to 547K Regular Army Force

	2009 Baseline (1)	980K Option			920K Option		
		980K Baseline (2)	Raw Shortfall (3)	AC Capacity to Replace Shortfall (4)	920K Baseline (5)	Raw Shortfall (6)	AC Capacity to Replace Shortfall (7)
Regular Army	547.4	450	(97.4)	97.4	420	(127.4)	127.4
ARNG	358.2	335	(15.0)	5.6	315	(35.0)	13.2
USAR	206.0	195	(11.0)	4.1	185	(21.0)	7.9
Wartime allowance				11.5			11.5
Total	1,111.6	980	(123.4)	118.7	920	(183.4)	160.0
Required AC end strength				568.7			580.0

SOURCES: McHugh and Odierno, 2014; Arroyo Center analysis.

NOTE: All values are in thousands. The 2009 baseline for the ARNG includes 8.2K of TTHS. Thus, the reduction to 335K only reduces 15K of force structure, or operational capacity.

mobilized was supposed be to 1:5.[4] In reality, these goals were not met throughout most of the conflicts in Iraq and Afghanistan. As we described in Chapter Two, Regular Army units generally deployed at a 1:1 BOG:Dwell ratio until late in the conflict. Initially, RC units were operating at a MOB:Dwell ratio of about 1:3, incurring considerable congressional scrutiny and proposals to impose the 1:5 ratio through statute. Congressional scrutiny diminished as MOB:Dwell rates declined to a little more than 1:4 during FY 2012 (FM&C, 2013, p. 5). We thus use the cyclic rotation rates the Army was actually able to sustain (1:1 for the Regular Army and 1:4 for the RC) as the basis for our estimates. Given the lower rotation rates of the USAR and ARNG, the AC capacity needed to make up for the RC shortfalls (columns 4 and 7 of Table 3.2) is less than the raw shortfall (columns 3 and 6 of Table 3.2).

In estimating required capacity, we also took into account TTHS, which, on average, includes about 13 percent of the Regular Army force; neither the USAR nor the ARNG is authorized a TTHS account. In addition, we included two other constraints on the analysis: time for overlap and, for RC forces, postmobilization training. We assumed a month of overlap for all units and two months of postmobilization training for RC units. Depending on their size and complexity, RC units required a certain amount of postmobilization training, ranging in duration from two to three

---

[4] The 2006 Army Posture Statement attributes a BOG:Dwell goal of 1:3 for the AC and 1:5 for the RC to a July 9, 2003 memorandum from the Secretary of Defense (see Harvey and Schoomaker, 2006, "Addendum E: Army Force Generation Model—ARFORGEN"). See also AR 525-2, 2011, p. 2, and DoD Directive 1235.10, 2011.

months, to attain specified levels of proficiency to deploy to theater (Klimas et al., 2014, p. 3). Both overlap and postmobilization training requirements increase the number of soldiers required to meet a persistent operational commitment. Finally, the totals also include a "wartime allowance" of 10,000 soldiers (plus a 13-percent allowance for TTHS). Previously, the Army has noted that such "temporary growth has improved the fill of priority units, reduced personnel turbulence and improved the Army Force Generation (ARFORGEN) unit manning with no additional structure growth" (FM&C, 2011a, p. 9).

Table 3.2 shows that the AC capacities needed to replace the shortfall are therefore 118,700 for the 980K option and 160,000 for the 920K option. Added to the starting AC sizes of 450,000 and 420,000, respectively, this yields a total requirement of 568,700 for the 980K option and 580,000 for the 920K option.

Conclusion

As Table 3.2 indicates, the Regular Army would have to expand by almost 120,000 soldiers to produce the same operational capacity as the 2009 baseline under the 980K scenario. Under the 920K scenario, the Regular Army would have to expand by 160,000 soldiers. Both options are somewhat larger than the peak strength of the Army at the conclusion of the Grow the Army initiative, but both options must compensate for the substantial numbers of reservists no longer available under these options.

This analysis calculated the raw, aggregate numbers of soldiers available to meet operational demands under some simplifying assumptions. It assumed that all soldiers not otherwise committed to another mission are available for deployment and that soldiers are essentially fungible across Army components, career management fields (CMFs), and unit types. That is, the analysis assumed that an RC soldier trained as an air defender can be employed as a military policeman with little degradation of operational effectiveness, compared with a Regular Army soldier fully trained in that CMF. In reality, soldiers are frequently nondeployable for various reasons (Arnold et al., 2011). Moreover, neither unit types nor soldiers' CMFs are perfectly fungible. Therefore, the Army would not be able to generate the full capacity predicted by these analyses. These limitations, however, apply equally to the initial condition and to the objective state. Thus, while these calculations do not fully reflect reality, they incorporate the necessary features of that reality sufficiently to enable relevant analysis. In Chapter Four, we explain the process we used to assess the Army's ability to recruit these soldiers within the target time frame.

Conceptual Framework and Policy Options for Regeneration

We developed a simple conceptual model of AC regeneration, which starts with initial end strength and seeks to satisfy the target end strength after a certain number of years, using various policy levers to affect flows of soldiers. Figure 4.1 illustrates the flows of soldiers that influence the size and composition of the force between any two years. We begin with a particular force size in Year 1. Soldiers flow into the AC from accessions, and flow out of the AC as a result of separations. The Army can influence flows into and out of the AC using accession and retention policies, respectively. In addition, there are flows within the AC, as soldiers are promoted to higher grades. Promotion policies will affect how quickly soldiers advance in rank and affect continuation rates, which tend to differ by grade and by time in service. The composition of the deployable force (including both AC and RC) is also influenced by the extent to which

Figure 4.1
Conceptual Model of Active Component Regeneration

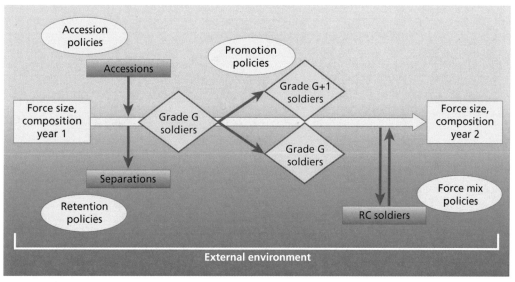

RAND RR1637-4.1

the RC is mobilized or demobilized. Such external factors as the economy and the job market also influence flows, albeit indirectly.

Both the 920K and 980K regeneration scenarios aim to meet targets for deployable AC end strength within five years. Figure 4.2 illustrates the timeline we modeled. We assumed that the conflict starts, and the need for regeneration is recognized, at the start of Year 1. Existing soldiers are immediately trained and deployed, but experience suggests that it will take time to obtain authorization for increased troop levels and to execute that plan. For example, even if troop authorizations were immediately increased, it would take several months to increase the number of recruiters and to train them, to obtain additional advertising slots and see results from the increased advertising efforts. Although higher enlistment and reenlistment bonuses could be implemented more quickly, it might also be several months before they took effect. Therefore, although existing troops are trained and deployed during the first year, we assumed that the major increases in accessions and retention are seen during Years 2, 3, and 4.

Although the surge in recruiting begins in Year 2, the new accessions must complete individual and unit training after being brought onboard. Training an individual soldier takes only about six months, but we made the simplifying assumption that it would take approximately one year from the time the Army starts trying to recruit the first soldier until the time the last soldier exits training and reaches his or her unit. The recruiting and training timeline is further complicated by the fact that enlisted soldiers are more likely to be available during the summer, immediately after high school

Figure 4.2
Conceptual Timeline of Active Component Regeneration

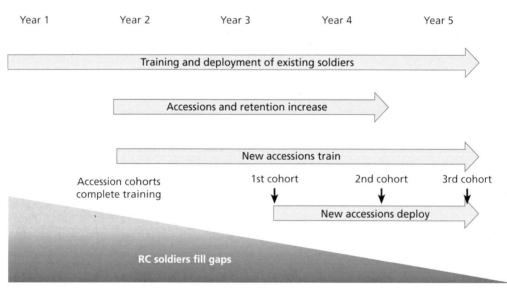

graduation. We therefore assumed that the first new "surge" cohort is onboard and has completed individual training by the end of Year 2. These individuals then train with their units during the following year and can be deployed at the end of Year 3.

We also assumed that the RC would fill the gap between demand for deployable troops and the number of AC troops available at a 1:1 BOG:Dwell ratio during this ramp-up period. During the first three years, as the Army prepares to ramp up accessions and trains new soldiers, the RC is assumed to cover the entire shortfall. RC participation then ratchets down as the new AC accessions are deployed.

Policy Options for Regeneration

In implementing the conceptual model, we examined a combination of accession, retention, and force management policy options for meeting the regeneration targets. Although a number of prior studies have considered one or more of these options, our goals were a holistic examination of how such policies might interact and an examination of the risks and costs associated with different policy combinations. We focused our efforts on enlisted soldiers, since the enlisted force has historically made up about 80 percent of AC end strength. The target enlisted AC end strength is approximately 455,000 under the 980K scenario or 464,000 under the 920K scenario.[1]

We implemented the conceptual model in three steps. First, we identified sets of accession policies that could be used to meet or approach the regeneration targets, given a reasonable cost and recruit quality mix (the share of soldiers considered "high quality," as defined by having high school diplomas and scoring at or above the median in the AFQT). Second, we identified retention and promotion policies that could be used to shape the resulting AC force. Third, we estimated the extent to which the RC could be mobilized to reach targets. The order of implementation is not meant to suggest that certain policies should be implemented before others, but was chosen as the most effective way of bringing together the various policy levers.

We also examined how the policy options would change under different external conditions. The main condition that we modeled quantitatively was prevailing economic conditions; as the civilian economy improves, the Army typically finds it harder to recruit and retain soldiers. However, as noted in Chapter Two, a variety of other conditions, such as public sentiment and propensity to serve, play a role in recruiting and retention. While some of these factors are harder to quantify, we explored scenarios that accounted for differences in observed retention rates during OEF/OIF.

The following subsections describe the three steps of model implementation in more detail.

---

[1] Enlisted targets are given by multiplying the Regular Army end strength targets from Table 3.2 by 0.8.

Accessions

We used the analytical model developed in Orvis et al., 2016, to estimate the maximum number of BCTs plus enablers and other supporting units/personnel that could be regenerated over the course of three years of "surge" recruiting.

The basic framework of the model is summarized here; Orvis et al., 2016, provides further details. The number of non–prior-service, high-quality contracts is assumed to be a function of the number of recruiters; spending on TV advertising; the size of the enlistment bonus paid to incentive-eligible, high-quality recruits; the proportion of high-quality recruits who are eligible for an enlistment bonus; and the unemployment rate. Model parameters, which govern how each of these factors affects the number of high-quality contracts, are calibrated to previous studies of recruiting. The number of lower-quality contracts is based on the predicted level of high-quality recruits and the assumed quality mix of enlistments.[2] The total number of contracts also is determined by the rate of enlistment waivers and prior-service accessions permitted under the enlistment eligibility policies being modeled.[3] The model then converts the predicted number of contracts into accessions by accounting for loss rates and patterns of accession from the DEP.

The model takes the number of recruiters and the unemployment rate as given. Additional inputs include the size of the entry DEP, the required share of high-quality accessions, the percentage of accessions that will be granted waivers, and the number of prior-service accessions. It then attempts to meet the user-specified enlisted force accession target at minimum cost by varying spending on TV advertising and enlistment bonuses.

The levels of TV advertising and enlistment bonuses are constrained in any given year based on the marginal cost of additional enlistment contracts—that is, the additional cost required to sign one additional contract—and on past Army policies. TV advertising costs are constrained to be no more than $192 million per year because the marginal cost becomes excessive beyond this amount. Enlistment bonus amounts for high-quality and lower-quality recruits are capped and floored based on the maximum and minimum observed, average per-recruit values in FYs 2006–2012. We do not cap the *total* amount of enlistment incentives that can be paid in any year but rather the average amount that can be paid to any individual recruit.

---

[2] We did not directly model lower-quality enlistments because there is little evidence on how recruiting policies affect such enlistees.

[3] Bonus costs increase to account for the need to compensate additional waivered non–prior-service and prior-service recruits (separate bonus value and eligibility rates are used for prior-service recruits, based on past Army practice and values). TV advertising costs and the number of recruiters do not increase because the loosened eligibility policies are treated as producing more return on these investments.

Applying the Accession Model to Regeneration

The model was originally designed to run for only a single year at a time; in this section, we describe how we used it to estimate BCT generation over the course of three years. The three years of the accession modeling component correspond to Years 2–4 in the conceptual regeneration framework (Figure 4.2).

We began by attempting to meet the baseline number of accessions in Year 2—that is, the number of accessions that would be needed, in the absence of regeneration, simply to replace soldiers who separate. For the purpose of this model, we assumed the baseline accession requirement at the 450K end strength to be 63,000,[4] then scaled this requirement down for the 420K end strength: 63,000 × (420/450) = 58,800. The entry DEP in Year 2 is assumed to be 25 percent of this baseline accession requirement. We set the number of recruiters and the enlistment eligibility policies (percentage of high-quality accessions, enlistment waivers, and number of prior-service recruits) to the desired levels for a given set of model runs.

After the model identified the combination of TV advertising and enlistment incentives that would meet the baseline accession requirement in Year 2 at a minimum cost, we attempted to rebuild one additional BCT in Year 2. The estimated number of additional accessions needed to rebuild one BCT was 16,364, including enabler unit accessions, TTHS, wartime allowance, table of distribution and allowances, and other supporting personnel, and replacement of within-recruiting-year losses of the new, regeneration recruits. In the remainder of this discussion, "BCT" refers to all these requirements.

As with the baseline accession requirement, the model attempted to build one additional BCT in Year 2, minimizing cost by varying TV advertising costs and enlistment bonuses and by keeping total TV advertising spending and average per-recruit bonuses under the ceilings discussed earlier. If the accession requirement for one additional BCT in Year 2 could be met, we then attempted to build a second BCT during the year. We repeated this exercise until we were unsuccessful in building another BCT because we had reached our advertising and bonus spending limits for a single year.

For Year 3, we began with the baseline accession requirement plus the shortfall in accessions needed to complete the BCT partially regenerated in the preceding year. We used the Exit DEP from the preceding year as the Entry DEP for the current year. The model accounted for DEP losses during a year. We also accounted for attrition from the BCTs already regenerated. These rates were based on historical data,[5] which indicated a loss rate of about 18 percent over the first year of service, 9 percent over the

---

[4] Based on RAND Arroyo Center discussions with Army G-1, which indicated that 63,000 would be the upper bound of a limited range in the estimated accession requirement considering possible variations in other factors, such as retention and attrition.

[5] Based on RAND Arroyo Center calculations using Total Army Personnel Database (TAPDB) data from 2000 onward.

second, 4.5 percent over the third, and 2.25 percent over the fourth (for a total of just under 34 percent over four years).

We calculated losses from attrition and included those in the newly calculated accession requirement as follows. We assumed that the average recruit accessed during the middle of each year of the regeneration process. For example, we assumed that the first cohort of soldiers brought in during the regeneration process would enter in the middle of Year 2 of the regeneration timeline. Since loss rates are about 18 percent over the first year of service, this leads to 9-percent replacements during Year 2 (built into the original accession requirement for that year) and another 9 percent to be replaced in Year 3. In addition, about one-half of the second-year-of-service loss rate (9 percent) would be experienced by the end of Year 3, adding another 4.5 percent that needed to be replaced during that year, for a total of 13.5 percent during Year 3. In Year 4 (the third year of unit regeneration), the remaining 4.5 percent of second-year-of-service losses plus the first half of third-year-of-service losses, 2.25 percent, needed to be replaced, for a total of 6.75 percent.[6] An analogous procedure was applied to recruits for BCTs regenerated in Year 3, yielding a replacement rate for these recruits of 13.5 percent in Year 4.[7]

Model Conditions

We ran the model under several different sets of starting conditions, which are summarized in Table 4.1. We estimated the model under three recruiting conditions: favorable, average, and unfavorable. Unfavorable recruiting conditions are characterized by low unemployment rates, because greater demand for civilian labor results in fewer potential recruits; conversely, favorable recruiting conditions are characterized by high unemployment rates. We assumed unemployment rates of 5.0, 6.5, and 8.0 percent for unfavorable, average, and favorable recruiting conditions, respectively. The *unfavorable* recruiting period reflects conditions during FYs 2006–2008, just before the Great Recession, when the monthly, seasonally adjusted civilian unemployment rate ranged from 4.4 to 6.1 percent. The *favorable* recruiting period reflects conditions during FYs 2010–2012, when the civilian unemployment rate ranged from 7.8 to 10 percent.[8]

---

[6] These rates are based on average observed attrition rates for incoming cohorts.

[7] This is based on the point, noted earlier, that within-recruiting-year losses are already included in the recruiting requirement for the year as part of meeting end strength. We did not compound the loss rates by applying attrition rates to soldiers who are recruited to replace those lost in the basic attrition calculations. For the example described here, we estimated the difference to be about 240 soldiers had the loss rates been compounded, with no effect on the total number of BCTs regenerated.

[8] We limited the unemployment rate used for favorable recruiting conditions to 8 percent instead of the exceptionally large values it reached during the Great Recession to keep it more in line with historical fluctuations and unfavorable to favorable recruiting cycles.

Table 4.1
Accession Scenarios and Inputs

Input	Scenario 1		Scenario 2	
AC end strength	420,000		450,000	
Baseline accession requirement	58,800		63,000	
Accessions needed for 1 BCT[a]	16,364		16,364	
Unemployment rates (%)				
Favorable	8.0		8.0	
Average	6.5		6.5	
Unfavorable	5.0		5.0	
Initial entry DEP (%)[b]	25		25	
Recruiters (number)				
Low	5,433		5,821	
High	6,011		6,440	
Enlistment Eligibility	**Greater**	**Lesser**	**Greater**	**Lesser**
High quality (%)	45	55	45	55
Prior service (number)	10,000	0	10,000	0
Waivers (%)	20	10	20	10

[a] Including enablers and other supporting units.

[b] Represents entry DEP in Year 2 of the conceptual framework and is a percentage of the baseline accession requirement.

The unemployment rate of 6.5 percent was selected to represent *normal* recruiting conditions in the middle of these two bounds.[9]

For each recruiting condition, we varied the number of recruiters who are assigned to recruiting stations and whose primary duties are to recruit youth into the Army. We considered two values for number of recruiters depending on the accession requirement. At a Total Army force size of 980K (an AC of 450K), we used 5,821 and 6,440 On-Production Regular Army (OPRA) "foxhole" recruiters.[10] To obtain the OPRA recruiter numbers to use for a Total Army force size of 920K (an AC of 420K), we

---

[9]   The civilian unemployment rates reflect monthly, seasonally adjusted figures from BLS, undated b.

[10]   The number for OPRA recruiters is the one that has been used in the recruiting research literature. The Army currently refers to Required Recruiting Force (RRF) on-production recruiters. There are about 0.92 OPRA foxhole recruiters for every RRF on-production recruiter. We selected the lower, baseline number (5,821) in keeping with the number of OPRA foxhole recruiters the Army was using in mid-FY 2013, at the time of initial model construction. We selected the higher recruiter number (6,440 recruiters) as the number of OPRA foxhole recruiters to represent the available 7,000 RRF recruiters, based on information from the Army on the metric it currently uses.

scaled the inputs we used for an AC of 450K by 420/450, resulting in values of 5,433 recruiters and 6,011 recruiters.[11]

In addition to the values we used for recruiters, we considered two enlistment eligibility policy scenarios that affected the number and quality of recruits. These scenarios fall generally under greater (that is, less restrictive thus more people eligible) and lesser (that is, more restrictive thus fewer people eligible) eligibility. The greater eligibility scenario is based on average recruit characteristics observed during the difficult FY 2006–2008 recruiting period, and is characterized by 45 percent high-quality accessions, 20 percent non–prior-service accessions with enlistment waivers, and 10,000 prior-service accessions. The lesser eligibility scenario is based on average recruit characteristics observed during the favorable FY 2010–2012 recruiting period, and is characterized by 55 percent high-quality accessions, 10 percent non–prior-service accessions with enlistment waivers, and no prior-service accessions.[12]

Example Results

To demonstrate how the model was used to assess regeneration potential, Table 4.2 presents our estimates for average recruiting conditions under an AC end strength of 450K (Total Army force size of 980K), with the lower number of recruiters, and with lesser enlistment eligibility policies (which are more consistent with Army preferences). Column 1 shows the recruiting goals and outcomes for the baseline accessions needed to maintain force size in Year 2 of the four-year period (year 1 of increased accessions). Column 2 then shows the results of increasing the baseline accession level to attempt to build one BCT in Year 2. In Year 2, we started at an initial entry DEP of 25 percent of 63,000 (15,750). In attempting to regenerate a BCT, we spent the maximum amounts on TV advertising and bonuses that were consistent with past Army policies.[13] This resulted in only partial regeneration of a BCT, which fell 3,123 accessions short of being complete for Year 2.

For Year 3, we adjusted the baseline accession requirement of 63,000 upward by accounting for the shortfall of 3,123 and attrition of 1,787. The attrition replacement of 1,787 was calculated by applying the attrition rate of 13.5 percent to soldiers regenerated during Year 2, i.e., to the difference between the total accessions of 76,241 and the baseline accession requirement of 63,000. The attrition replacement for the baseline (nonregeneration) portion is built into the 63,000 requirement. We also adjusted the Year 3 entry DEP by taking the end-of-year DEP from the previous year (22,141). This completed regeneration of the first BCT. Next, we attempted to regenerate a second

---

[11] We arrived at the number of recruiters for a 420K AC by scaling down the number of recruiters for a 450K AC proportionally (e.g., 6,440 × 420/450 = 6,011).

[12] Data on recruiting characteristics are based on information in the Regular Army (RA) Analyst database.

[13] As noted above, the ceilings were based on the marginal cost of additional enlistment contracts and on past Army policies. The ceilings apply to total TV advertising and to per-recipient average enlistment bonus amounts.

Table 4.2
Example of Regeneration Results for 450K, Average Recruiting Conditions, 5,821 OPRA Foxhole Recruiters, and Lesser Enlistment Eligibility Policies

	Year 2		Year 3		Year 4	
	(1)	(2)	(3)	(4)	(5)	(6)
Accession goal	63,000	79,364	67,910	84,274	68,580	84,944
Entry DEP (entry pool)	15,750	15,750	22,141	22,141	22,203	22,203
Accessions	63,000	76,241	67,910	81,897	68,580	81,951
Shortfall	N/A	3,123	N/A	2,377	N/A	2,993
Cost[a]						
Recruiters	686,878	686,878	686,878	686,878	686,878	686,878
TV advertising	180,575	192,000	180,666	192,000	181,163	192,000
Total bonus	244,765	1,005,896	227,118	1,012,808	254,708	1,013,550
Total cost	1,112,218	1,884,774	1,094,662	1,891,686	1,122,749	1,892,428
End-of-year DEP	23,678	22,141	23,645	22,203	23,671	22,204

[a] In thousands of 2015 dollars.

BCT, which, as for Year 2, added 16,364 to the accession requirement. This again resulted in only partial regeneration of a BCT, which was 2,377 accessions short of being complete for Year 3.

For Year 4, we again adjusted the baseline accession requirement of 63,000 upward, in this case by accounting for a shortfall of 2,377 and attrition of 3,203. That attrition value was calculated by applying the attrition rate of 6.75 percent to soldiers regenerated during Year 2 and a rate of 13.5 percent to soldiers regenerated during Year 3. We also adjusted the Year 4 entry DEP by taking the end-of-year DEP from the previous year (22,203). This again completed regeneration of the BCT started in the prior year (Year 3) plus partial regeneration of a second BCT. The shortfall for the second BCT in Year 4 was 2,993 for maximum spending on advertising and enlistment bonuses that was consistent with past Army policies.

In the end, under this scenario, the iterations over the three-year period regenerated two complete BCTs, plus their enablers and other supporting units and soldiers. The estimated total cost for the complete BCTs is the sum of the total cost entries in columns 2 and 4 for Years 2 and 3, respectively, and in column 5 for Year 4.[14]

We applied this set of steps to each of the different input conditions shown in Table 4.1 for both the 920K and 980K scenarios. In addition, as noted in Chapter Two, we considered a historical accession scenario that assumed that the Army can recruit the maximum number of accessions seen since 9/11: 80,000 per year. Approximately 80,000 accessions were achieved in FY 2002 and in FYs 2006–2008, during the Grow

---

[14] The cost in column 6 is associated with a BCT that is only partially regenerated in Year 4.

the Army campaign. This figure should not be seen as an upper limit on the number of recruits that the Army could bring in: In FY 2002, recruiting resources were fairly low, and eligibility standards were relatively high (lesser eligibility); during FYs 2006–2008, a civilian unemployment rate that fell below 5 percent created very difficult recruiting conditions. In addition, in both cases, the accession numbers reflected the recruiting goals that the Army was trying to reach. Nonetheless, we explore the 80,000 accession scenario as a benchmark for the maximum number of recruits that has actually been brought in during recent years.

Retention

The retention analysis takes as inputs the numbers of soldiers entering the Army that are outputs from the accession model above. It uses a Markov-chain inventory projection model (IPM) to project forward how these soldiers progress through the system based on their rank and years of service (YOS) and the resulting shape of the AC. The following are the key inputs to the IPM:

- the initial (starting) inventory of soldiers
- the objective (final) inventory of soldiers
- the continuation rates—the proportion of soldiers in a given grade and YOS that will continue to serve in the following year (also known as a survival rate)
- promotion eligibility ranges—enlisted soldiers are eligible for promotion after accruing a certain number of months of service, as determined by Army policy.

The last item of promotion eligibility ranges reflects policy variables that can be manipulated in the IPM, reflecting, for example, a need to promote more people earlier to meet requirements for more NCOs in an expanding force. Continuation rates can similarly be manipulated to reflect estimated effects of retention incentives, or policies (such as stop loss) designed to enable more expansion.

Expanding the population of enlisted soldiers differs from expanding a civilian organization insofar as leaders must be exclusively "grown" from within; lateral entry at higher grades is uncommon. Thus, at any given time, the size (and shape) of the Army is a function of its size (and shape) at some point in the past, the number of new recruits since then, and the number of soldiers who have left. When examining progression through the Army's ranks, a similar construct holds: The number of soldiers at a particular grade is a function of the number of soldiers in that grade at some point in the past; the number promoted into that grade from the next lower grade; and the number of soldiers who leave that grade, either by promotion, or by separation from the Army.

In light of these facts, while the accession model above asks how many new recruits and soldiers the Army can bring in the door, the retention model is essentially a stock-and-flow model of cohorts that asks: How do these recruits progress through the Army's ranks, and what is the resulting shape of the force?

Available Policy Levers for Retention and Associated Guidelines

The primary tools available to help the Army retain its size and determine its shape include increasing promotion rates, decreasing time-in-service requirements for promotion, and increasing retention incentives to persuade soldiers to remain. The Army regularly adjusts these factors to suit its needs in maintaining an adequate inventory of soldiers at higher grades in the face of changing circumstances. Accordingly, the IPM can manipulate these factors to adjust the flow of soldier cohorts through the system to achieve the objective ending inventory of soldiers.

Although allowed to adjust such factors, the Army has myriad policies and historical precedent governing how soldiers progress through the system and at what speed, as well as what sorts of retention incentives it may offer and to what types of soldiers. The IPM takes these guidelines into account in setting its parameters.

Promotions

AR 600-8-19, 2011, is the main policy document that governs how and how quickly enlisted soldiers progress through the ranks. For example, Army regulation typically requires that soldiers should acquire seven YOS before advancing to the rank of staff sergeant. Though there are exceptions—in some instances commanders can use waivers to promote exceptional soldiers early or in certain careers or specialty populations—the majority of the enlisted force progresses according to the main policies outlined by the Army.

Accordingly, the IPM incorporates these major promotion rules as inputs into its stock-and-flow model but allows soldiers to be promoted earlier if needed to shape the force. Promotions from E-1 through E-4 are "decentralized," and soldiers are typically promoted in lockstep, unless there is some reason for deviation from the normal. The model therefore assumes that nearly all soldiers are promoted to E-4 by the end of two YOS.

Promotions to E-5 and E-6 are "semicentralized"; soldiers are typically promoted to E-5 by four YOS but can be considered for promotion after as few as 18 months in service. Similarly, soldiers are typically promoted to E-6 after seven YOS but can be considered after only four YOS. We experimented with allowing the model to promote soldiers after the minimum necessary time in service, but this resulted in unrealistic progressions of rank for many soldiers. Thus, the model allows promotion to E-5 by three YOS and E-6 by five YOS, which are earlier than the typical YOS at promotion but not the earliest service policy allows. Promotions to E-7, E-8, and E-9 are determined by centralized selection boards, and AR 600-8-19 does not provide specific time-in-service criteria for these promotions. In keeping with observed historical promotion patterns, the model allowed promotion to E-7 as early as ten YOS, to E-8 as early as 14 YOS, and to E-9 as early as 18 YOS. In Chapter Five, we discuss the implications for average NCO experience when the Army promotes large numbers of them earlier.

We also constrained the model's allowable rates of promotions and accessions to ensure that the shape of the AC force does not fluctuate dramatically from year to year. Specifically, we set the model not to allow the share of soldiers in any grade to increase or decrease by more than 10 percent in any given year. This does not mean that the *number* of soldiers could not increase by more than 10 percent in any grade, but rather that the overall shape of the AC—the *relative* numbers of soldiers in each grade—must remain fairly stable.

These constraints on promotion and on force shaping mean that, in certain cases, the accession model produces more new recruits than the IPM can absorb. This occurs almost exclusively in cases in which the required recruit quality mix was lowered, and the accession model thus estimated that more than 100,000 new recruits could be produced each year. Therefore, in no case did these constraints cause us to estimate a shortfall when removing the constraints would have allowed us to meet the targets. In these cases, we assumed that not all recruits were accessed (and lowered the estimated accession costs accordingly).

Incentives for Retention

Another key input to the IPM is the continuation rates for soldiers at given ranks and experience levels. Historically, retention incentives have been one of the Army's main policy instruments for managing its force; more soldiers accept the incentive and remain in service, raising continuation rates. AR 601-280, 2011, is the most recent document outlining bonus eligibility and guidelines for how the Army can use financial incentives to encourage AC reenlistments and retention. AR 601-280, Ch. 5, Sec. I, lists all the eligibility requirements, which differ slightly based on YOS and other factors. However, two basic requirements hold in all cases: (1) The soldier must be active-duty Army, and (2) the soldier must reenlist for a minimum of three years.

Use of financial incentives to retain its workforce is not new for the Army or any of the military services (Asch, Heaton, et al., 2010). The Army has used bonus incentives to shape its personnel structure since the 18th century (Office of the Under Secretary of Defense [Personnel and Readiness], 2005, p. 611; Asch, Heaton, et al., 2010, p. 43). Personnel management and reenlistment bonuses became especially important after the U.S. military transitioned to the All-Volunteer Force in 1973 (Asch, Heaton, et al., 2010). During the following year, in 1974, Congress established the official Selective Reenlistment Bonus (SRB) program, which largely remained in place with only minor adjustments until 2007.

In 2007 the Army's SRB program underwent significant changes, and the Army began a derivative program, the Enhanced SRB. The Enhanced SRB is the major financial program the Army uses to encourage soldiers to reenlist and constitutes one of the most direct and commonly used policy levers available to the Army for managing its active-duty enlisted workforce (Asch, Heaton, et al., 2010). The Enhanced SRB specifies different bonus amounts depending on the soldier's YOS, MOS, rank, and amount of additional obligated service (AOS), which are announced by military personnel messages. Although AR 601-280 requires a minimum reenlistment of three years, the military personnel messages typically list bonus amounts for AOS as low as 12 months.

For this modeling exercise, we calculated average continuation rates by grade and YOS through an analysis of TAPDB records from 2003 through 2012 as the input for our baseline scenario. We then adjusted these continuation rates based on estimates from the literature on the effects of reenlistment bonuses on retention. That is, we turned the policy lever of increased reenlistment bonuses "on" and adjusted overall continuation rates upward by an estimated effect size. Later, we will discuss how the bonuses were applied in more detail.

It is also worth noting that, starting in 2018, the Army will phase in a pension plan that includes elements of a defined contribution plan, pays a bonus to soldiers who have 12 YOS and agree to remain for four more years, and maintains retirement pay for those who serve 20 YOS. Previous research indicates that changes to retirement plans can substantially influence retention throughout a soldier's career.[15]

Retention Scenarios

We considered five scenarios in the retention model:

- a baseline case that takes average continuation rates from 2003–2012 from TAPDB data
- a "low" estimated effect from introducing reenlistment bonuses that increases the baseline continuation rates somewhat
- a corresponding "high" estimate from reenlistment bonuses that has a larger effect on the baseline continuation rates
- an upper-bound estimate that takes the average continuation rates drawn from years 2007–2009
- a corresponding lower-bound estimate that takes average continuation rates from years 2003–2005.

The last two bounded estimated continuation rates were meant to serve as a sensitivity analysis and were chosen to comprise recent history's highest and lowest recorded continuation rates. They necessarily encompass the economic and political climates of these years, the progression of the OIF/OEF wars, and the Army's own retention policies, including reenlistment incentives and the use of stop-loss policies, which were implemented to varying degrees during this time, with a peak of 20.5 percent of soldiers subject to stop loss in FY 2005 (Simon and Warner, 2010).

Rather than try to model separately the effects of these factors, we treated the continuation rates from FYs 2007–2009 and FYs 2003–2005, respectively, as the best-

---

[15] See, for example, Asch, Mattock, and Hosek, 2015.

and worst-case scenarios. For example, the lower-bound rates represent a time of good economic performance and low retention bonuses.[16] In contrast, the upper-bound rates can be taken as representative of a time when the poor performance of the civilian economy (at least during the latter half of the period), high retention bonuses, and the use of stop loss were all contributing to keeping retention rates high. In Chapter Five, we show a historical scenario that reflects the highest number of accessions and the upper bound of retention seen during the past 15 years. We refer to this as the Grow the Army scenario because it assumes accessions of 80,000 new recruits per year, coupled with the high continuation rates seen during FYs 2007–2009, both of which are representative of the Grow the Army campaign.

The remaining three scenarios manipulate continuation rates as inputs to the IPM by turning an SRB policy on or off. As discussed above, a continuation (survival) rate is the proportion of soldiers in a given grade and with a given YOS who continue on to serve the following year. The inverse of this continuation rate is, by definition, an attrition rate: the rate at which soldiers leave the service. Most of those who leave the service do so at expiration of the term of service (ETS), that is, at the end of their original (or previous) active-duty service obligation. Some smaller fraction will attrit in the middle of their term of service, because of behavioral infractions or other problems. Reenlistment bonus incentives are meant to encourage soldiers who are nearing their ETS to sign on for another term. Accordingly, reenlistment bonuses can only be awarded to those soldiers eligible to reenlist, which has, since 2005, generally included soldiers within 24 months of their ETS. Thus, the effects of turning on a bonus lever will only affect those who can receive it.

The overall continuation rate of soldiers within a given rank and YOS cohort in a given year can be expressed as the weighted average of the reenlistment rate among those eligible to reenlist and the rate of nonattrition among those in the middle of their terms: that is, the share of soldiers eligible to reenlist (the share within 24 months of their ETS) multiplied by that cohort's reenlistment rate, plus the share ineligible to reenlist multiplied by the rate of nonattrition for midterm soldiers.

In applying the effects of an SRB policy to the IPM, for each rank and YOS cohort, we calculated the average continuation rate from FYs 2003–2012, the average share of soldiers who are within 24 months of their ETS, and the average reenlistment rates for those who are eligible to reenlist. We also observed the nonattrition rate of those who were ineligible. We then adjusted the overall continuation rates for each rank and YOS cohort by adding our estimated effects of an SRB policy to these underlying reenlistment rates, applied only to the share of those eligible to receive the bonuses. This means that the effects of SRBs are somewhat diluted when incorporated

---

[16] We note that FY 2005 forms part of our "lower bound" although stop loss actually peaked during this year. Other conditions—including a strong economy and the fact that the Army had not expanded recruiting payments or eligibility substantially at this point—overwhelmed the effects of the stop loss, resulting in relatively low retention during this year.

into the overall continuation rates because SRBs can affect only a fraction of soldiers in that cohort. Additionally, we adjusted the continuation rates only for soldiers with less than 15 YOS, since SRBs are generally not offered to soldiers with higher YOS. We drew our choice of the low and high estimated effects from an SRB policy in Asch, Heaton, et al., 2010, whose estimates align with other studies, which we summarize briefly below.

History of the SRB Program and Its Effects on Reenlistment and Retention

Given how important retention bonuses are to Army officials for shaping the enlisted force, many studies have estimated their effects on retention. However, in our search of the literature, all the published works estimating the effects of SRBs on reenlistments used data before 2008, meaning the analyses focused on effects of the SRB program before the advent of the current Enhanced SRB program.

The original SRB program that Congress established in 1974 awarded soldiers an amount based on the following equation: SRB = AOS × SRBM × MBP, where AOS is in years, SRBM is the SRB multiplier, and MBP is monthly basic pay (Asch, Heaton, et al., 2010). The SRBM was ultimately what the Army used to regulate bonus levels. In general, the multiplier was a number from 0 to 6 and was allowed to vary in half-unit increments (Asch, Heaton, et al., 2010, p. 44).[17] The literature on the elasticity of SRBs generally estimates the effect of a one-unit increase in the SRBM on the probability of reenlistment.

In 2007, the Army adjusted the reenlistment bonus program, establishing the Enhanced SRB program. The major difference between the Enhanced SRB program and the former SRB program is that the bonus amount for the Enhanced SRB program no longer depends on a multiplier (the SRBM). Instead, bonuses under the Enhanced SRB program are a function of MOS, rank, YOS, and the additional YOS a soldier signs up for, up to five years (Asch, Heaton, et al., 2010, p. 49).

To the best of our knowledge, no published studies have estimated the effects of this Enhanced SRB program on reenlistment since 2007. This has forced us to use an estimate based on the SRBM and meant that we had to update our elasticity estimate to a dollar-value effect. Although this leaves open the question of whether the fundamental relationship between bonus incentives and reenlistment may have changed since the policy change, Hansen and Wenger, 2002, found that differences in elasticity estimates across multiple studies and over many years for the effects of incentives on sailors are primarily the result of modeling differences and not of any changes in

---

[17] Specifically, Asch, Heaton, et al., 2010, p. 44, states that

> [l]egislation establishing the SRB program originally permitted SRB multipliers from 0 to 6. The law was amended in FY 1989 to permit multipliers of up to 10. See [Office of the Under Secretary of Defense (Personnel and Readiness), 2005, p. 625]. The Navy is the only service to have taken advantage of this increase in the maximum multiplier.

underlying responsiveness to pay. We assumed that a similar underlying relationship holds here.

We took our elasticity estimates from Asch, Heaton, et al., 2010, Table 7.6, which shows that, for zones A and B combined (where *zone* translates to a soldier's YOS), the SRBM's effect on the probability of reenlistment ranges from 3.5 to 5.6 percentage points, depending on whether the soldier is deployed. Asch, Heaton, et al., 2010 considered these to be the lower- and upper-bound estimates for the effects of SRBs because conditioning on deployment is likely to overstate the effects, while not doing so is likely to understate them. We chose the estimates that combine the effects for zones A and B because the changes in bonus policy after 2007 reduced disparities between zones A and B in terms of average bonus size (Asch, Heaton, et al., 2010). These estimates are in the range of historical estimates from the broader literature; Simon and Warner (2010) summarize several reviews of the literature from previous studies, as well as two additional studies not covered in those reviews, and note that the estimated effects of the SRBM typically range between 2 and 6 percentage points.

Our lower-bound scenario from turning on a higher SRB is that a one-unit increase in the SRBM leads to a 3.5-percentage-point increase in the probability of reenlistment. To put this into dollar terms, we took the most common reenlistment term of three years in the absence of an incentive (at a multiplier of zero), then added six additional months to this for a new reenlistment term of 3.5 years when the SRB is turned on (see Asch, Heaton, et al., 2010, Table 7.9, for those who go from an SRBM of zero to an SRBM of one in zone A). We multiplied this new reenlistment term by average monthly base pay of $2,351.40, the amount for a soldier at grade E-4 with more than four YOS in 2015 (Defense Finance and Accounting Service, 2016). This implies that a one-unit increase in the SRBM translates to an average increase in bonus size of $8,229.90 in 2015 dollars. In turn, taking Asch's lower-bound estimate, an $8,229.90 increase in bonus size leads to a 3.5-percentage-point increase in the probability of reenlistment, or every $1,000 additional in bonus size increases reenlistments by 0.43 percentage points. The Asch, Heaton, et al., 2010, upper-bound estimate that a one-unit increase in the SRBM leads to a 5.6-percentage-point increase in the probability of reenlistment translates into a 0.68-percentage-point increase in the probability of reenlistment for every $1,000 increase in bonus size.

History supports an average bonus increase of approximately $8,000. First, a one-unit increase in the multiplier matches historical patterns for manipulating bonus levels, particularly during the Grow the Army campaign (see Simon and Warner, 2010, Fig. 1, p. 512). Second, we compared the modeled increase with the actual dollar increases reported during this time. In FY 2001, the average SRB was $8,436 in nominal dollars, and 17,125 initial payments were given among the 64,982 soldiers who reenlisted. Thus, just over one-quarter of reenlisting soldiers received bonuses. During the Grow the Army period (FYs 2006–2008), the number of initial bonus payments rose dramatically (reaching a peak of 65,156 in FY 2006), as did the bonus amount (reaching $10,600 in FY 2006; $12,400 in FY 2007; and $13,600 by FY 2008, all in terms of nominal dollars). The proportion of soldiers receiving bonuses increased from roughly 20–25 percent during the early 2000s to a peak of 97 percent in FY 2006, falling to 50–60 percent in FYs 2007 and 2008.[18]

Given the increases not only in bonus amounts but also in the shares of reenlisting soldiers receiving bonuses, we viewed the SRB policy in terms of its *expected value*: that is, the amount of the bonus, multiplied by the probability of receiving it. In FY 2001, the expected value in nominal terms was $8,436 × (17,125/64,982) = $2,223. In FY 2006, the expected value in nominal terms was $10,600 × (65,156/67,307) = $10,261, an increase of approximately $8,040.[19]

We therefore assumed an average bonus increase of $8,230 (in keeping with the one-SRBM increase during Grow the Army) and calculated continuation rates given the low- and high-SRB effect sizes. Figure 4.3 summarizes the continuation rates that result from applying the low- and high-SRB effects, as well as the continuation rates from the baseline and upper- and lower-bound scenarios. Continuation rates for YOS 1 and 2 are close to 90 percent, and the sharp drop in continuation rates in YOS 3 and 4 is consistent with typical initial enlistment terms of 3–4 years. Continuation rates tend to rise among those who elect to reenlist after the first term, plateauing at around 95 percent by YOS 15. By construction, turning on the SRB increases continuation rates slightly at all YOS. For most YOS, the upper-bound rates are above the rates estimated using the high-SRB effects. However, the upper-bound rates are lower than most of the other rates for YOS 1 and 2, as a result of somewhat higher attrition among recruits during the FY 2007–2009 period. Similarly, the lower-bound rates are typically below the baseline rates. In any event, the average continuation rates do not differ significantly across scenarios, although even small differences can have notable effects after the IPM compounds the differences over time.

Force Mix

We also examined the extent to which the Army would have to draw on the RC to support the ongoing level of commitment envisioned in the scenarios. As shown in the timeline (Figure 4.2), we assumed that the major increases in AC accessions occur during years 2, 3, and 4 and that it takes approximately one year between when the Army starts trying to recruit an increased number of accessions and when the last new recruit completes individual training, plus an additional year for the new unit to which those additional recruits would be assigned to accomplish collective training. Thus, our modeling assumes that no substantial increment to deployable AC capacity will be

---

[18] Initial bonus payment amounts and number of bonuses are from FM&C budget estimates (various years). Total number of reenlistments is from Asch, Heaton, et al., 2010, Table 6.4.

[19] If the bonus values were converted into constant FY 2015 dollars, the increase would be approximately $9,089.

Figure 4.3
Average Continuation Rates for Each of Five Scenarios, by YOS

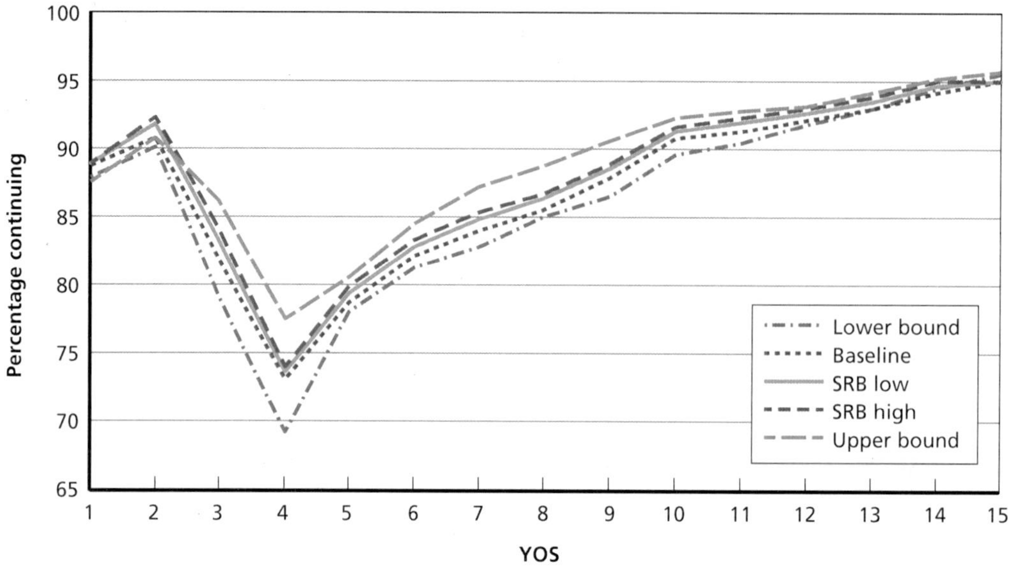

NOTE: Figure presents weighted averages for average continuation rate across all ranks in a given YOS.
RAND RR1637-4.3

available during the first three years of the conflict. Instead, reserve forces will have to operate at a higher deployment tempo to meet the deployment targets in the short run.

We made additional simplifying assumptions:

- AC and RC soldiers are perfectly substitutable.
- All RC units require two months of predeployment training.
- RC units require a one-month overlap for relief in place or transfer of authority.

To find the duration of the cycle, we divided the cumulative supply of RC operating force manpower available for deployment over a given period—an amount that differed according to the scenario—by the cumulative mobilization requirement for RC forces over that period. The mobilization requirement is significantly higher than the actual requirement for forces in theater because postmobilization training reduces the operational availability of RC forces. We subtracted one year—the nominal period of mobilization—from the resulting ratio—the length of the entire cycle—to find the dwell time: [20]

---

[20] We derived the cumulative mobilization requirement as follows:

- RC units are assumed to require two months of postmobilization training and (as with AC units) an additional month of overlap and so are actually available for operational employment for only 9 months of a

$$Dwell = \frac{\sum_{year=1}^{n} RC\ Operating\ Force\ Manpower_{year}}{\sum_{year=1}^{n} Mob.\ Reqt.\ for\ RC\ Operating\ Forces_{year}} - 1$$

Note that we estimated the *cumulative* MOB:Dwell ratio; for example, if the ratio is 1:3 in Year 5, the RC is not necessarily operating at 1:3 in Year 5, but *has been operating* at an average ratio of 1:3 over the previous five years. In reality, neither units nor soldiers are completely fungible, so the cumulative MOB:Dwell rate for soldiers in the required units will likely be higher.

Summary

We examined the Army's ability to regenerate active duty end strength given the different scenarios, starting conditions, external conditions, and policy levers discussed in this chapter. Table 4.3 summarizes the various assumptions and scenarios that we considered. Chapter Five presents the results of the modeling based on the methods presented here.

12-month mobilization cycle. Thus, the available operational capacity is 9/12 of the total mobilized force, and we found the mobilization requirement by dividing the required operational capacity by 9/12.

- To find the required operational capacity, we added the amount we originally planned to obtain from the RC at a 1:4 ratio to the shortfall in AC capacity. The original RC capacity is equal to total rotational capacity divided by 5 (5 = 1 + 4) × 9/12.
- To find the shortfall in AC operational capacity, we multiplied the shortfall in end strength by 11/24, the ratio of months available to overall cycle length in a two year cycle, assuming a 1:1 BOG:Dwell rate and 1 month overlap.

Table 4.3
Summary of Conditions and Assumptions Considered for Modeling Accessions, Retention, and Force Mix

Condition	Range of Options Considered
Initial end strength	
Scenario 1	980K (450K active)
Scenario 2	920K (420K active)
Regeneration objective	Capacity provided by 547.4K (2009 baseline for Army)
Shortfall to fill	
Scenario 1	118,700
Scenario 2	160,000
Regeneration time	5 years
AC BOG:Dwell rate	1:1
RC MOB:Dwell rate	1:4
Size of entry DEP	25 percent of baseline accession mission
Recruiting standards	
Lesser enlistment eligibility	55 percent high school diploma, Categories I–IIIA 10 percent waivers
Greater enlistment eligibility	45 percent high school diploma, Categories I–IIIA 20 percent waivers 10,000 prior-service recruits
Continuation rates	Average historical rates (2002–2012)
	Low rates (2003–2005)
	High rates (2007–2009)
Promotion rates	Based on combination of historical promotion rates, Army regulations, and limiting share of soldiers in each grade to change by <10 percent per year
Unemployment rate (%)	
Favorable recruiting conditions	8
Average recruiting conditions	6.5
Unfavorable recruiting conditions	5
Retention bonus size	Equivalent to one-unit increase in SRBM (approximately $8,200)
Retention bonus elasticity	
Low effect	One-unit increase in SRBM results in 3.5-percentage-point increase in retention
High effect	One-unit increase in SRBM results in 5.6-percentage-point increase in retention

CHAPTER FIVE

Modeling Results

This chapter summarizes key results from the combined accession and retention model, presenting the shortfall, that is, the difference between estimated AC end strength and target end strength under a variety of external conditions and using different policy options. We present shortfall in terms of enlisted AC end strength relative to the enlisted AC end strength target—463,000 for the 920K regeneration scenarios, and 454,000 for the 980K scenarios—at the end of Year 5.

Two points should be noted when interpreting these shortfalls. First, the shortfall shown for each year is based on the number of soldiers who can actually be deployed in that year, not the number of soldiers onboard. As discussed in Chapter Four, although accessions and retention increase starting in Year 2, the first cohort of surge accessions is not available to deploy sooner than the end of Year 3, while the third (last) cohort of surge accessions is not available to deploy until the end of Year 5. Second, the shortfalls are given in terms of the final end strength targets, implicitly assuming an immediate increase in demand to the target levels. If the buildup in demand is gradual, the shortfalls will be lower than those shown for Years 1–4.

We also examine the extent to which the RC would need to be mobilized to fill the estimated shortfalls. In this event, we explicitly consider the case in which there is an immediate need for the full strength of deployable forces and the case in which need starts from a base level and increases gradually to full capacity over time.

920K Scenario

Effects of Accessions and Retention Policies Under Average Conditions

We begin by estimating the potential for accession and retention policies to meet the required targets for deployed forces at the end of a five-year time frame, given average civilian employment conditions (that is, an unemployment rate of 6.5 percent). Figure 5.1 shows the estimated shortfall—the number of enlisted soldiers short of the target at the end of each year—and how this shortfall changes as the Army utilizes different policy levers for accessions and retention.

Figure 5.1
920K Scenario: Estimated Enlisted Shortfall Using Different Policy Levers

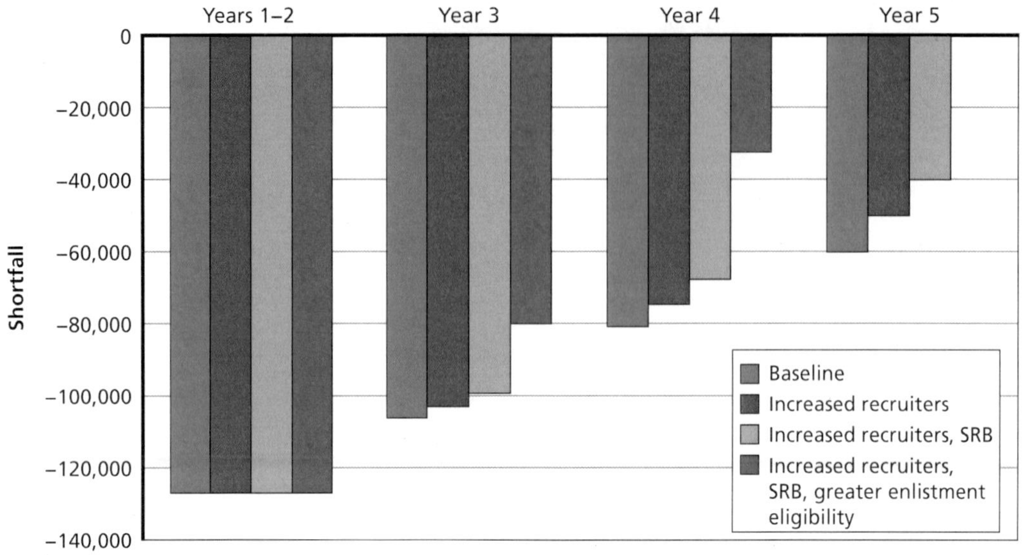

The baseline results in Figure 5.1 (blue bars) consider a case in which the number of recruiters is relatively low, and only accession policies—a combination of enlistment bonuses and TV advertising—are used to increase enlistment. The accession model suggests that, under this baseline scenario, the combination of enlistment bonuses and TV advertising could bring in approximately 77,000 recruits per year. At the end of Year 5, the size of the enlisted force is approximately 60,000 soldiers below the target of 463,000.

Additional accession and retention policy approaches can further decrease this shortfall. Increasing the number of recruiters (red bars) increases the estimated number of accessions to approximately 81,000 per year, and thus lowers the shortfall to approximately 50,000 by the end of Year 5. If the SRB is also increased as discussed in Chapter Four, we estimate that the increase in retention lowers the shortfall to just under 40,000 (green bars) by the end of Year 5. For the results in this chapter, we show the low-SRB effect (see Chapter Four). Appendix A shows that the results of assuming a high-SRB effect are similar.

While these policy approaches mitigate the gap, they do not close it. However, if optimal advertising and enlistment bonus policies, as well as more recruiters, are coupled with an increase in enlistment eligibility for new recruits (requiring that only 45 percent, rather than 55 percent, be high-quality recruits; allowing 10,000 prior-service accessions; and increasing waivers from 10 to 20 percent), the modeling results suggest that the shortfall can be eliminated by the end of Year 5. The reason is that allowing more lower-quality recruits increases recruiter productivity (in terms of the number of new recruits) substantially, so that accessions are estimated to reach more than 105,000 annually. While this accession level has not been seen in recent years, accessions of more than 120,000 were common before the drawdown in the 1980s.

Effects of Different External Conditions

In this subsection, we estimate the effects of different external factors on the Army's ability to regenerate the AC. For the remainder of the analyses pertaining to the 920K scenario, we focus on results that assume that the Army draws on all the available accession and retention policy levers—increased recruiters, TV advertising, enlistment bonuses, and reenlistment bonuses—but may or may not find it appropriate to lower the required recruit quality mix.

Figure 5.2 shows that if more stringent enlistment eligibility standards are used (i.e., a high-quality mix of enlistees is targeted), then favorable recruiting conditions (a high civilian unemployment rate) can lower the shortfall from 40,000 (under average conditions) to 26,000 (under favorable conditions) by the end of Year 5. In contrast, unfavorable recruiting conditions (a low civilian unemployment rate) can increase the shortfall to nearly 55,000.

Figure 5.3 shows the effects of allowing greater enlistment eligibility on estimated shortfalls under different external conditions. Even under unfavorable recruiting conditions, the Army could likely meet its target by the end of Year 5 if it were to accept more lower-quality recruits and prior-service accessions and to grant more waivers. The reason is that these policies increase the number of potential recruits so much that

Figure 5.2
920K Scenario: Estimated Enlisted Shortfall Under Different Conditions

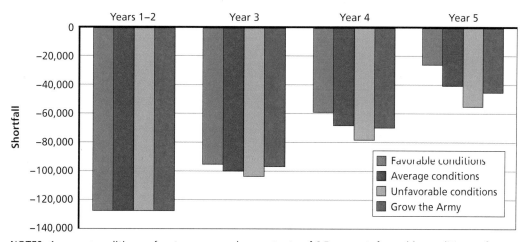

NOTES: *Average conditions* refers to an unemployment rate of 6.5 percent; *favorable conditions* refers to an unemployment rate of 8 percent; and *unfavorable conditions* refers to an unemployment rate of 5 percent. The Grow the Army scenario is based on 80,000 recruits per year and retention rates observed during FYs 2007–2009.

RAND *RR1637-5.2*

Figure 5.3
920K Scenario: Estimated Enlisted Shortfall Under Different Conditions, Greater Enlistment Eligibility

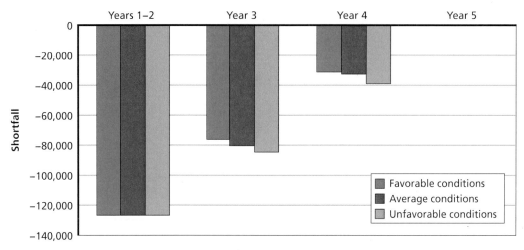

NOTES: *Average conditions* refers to an unemployment rate of 6.5 percent; *favorable conditions* refers to an unemployment rate of 8 percent; and *unfavorable conditions* refers to an unemployment rate of 5 percent.
RAND *RR1637-5.3*

there are more recruits than can be accommodated or are needed to meet targets. In contrast, Figure 5.2 showed that the Army would face substantially greater shortfalls under unfavorable conditions than under favorable conditions if it were to maintain lesser enlistment eligibility standards. These findings suggest that, if greater eligibility is acceptable, increasing enlistment eligibility can be used to reduce or eliminate shortfalls if the need for growth occurs during unfavorable conditions.

While we can model the effects of the civilian unemployment rate explicitly, a number of factors are more difficult to quantify, including public sentiment about the conflict and the overall propensity to enlist. In addition, while we model the major policy levers that the Army uses to recruit and retain soldiers, a number of other levers, such as stop loss, are not practical to model explicitly. We therefore show the results of a Grow the Army scenario, which assumes that the Army can access approximately 80,000 soldiers each year (in line with the numbers seen during the Grow the Army initiative) and that continuation rates are equal to those witnessed during this period (the upper-bound rates discussed in Chapter Four). As discussed in Chapter Four, these accession and retention figures do not represent upper limits but rather reflect a variety of economic, social, and political conditions that prevailed during the Grow the Army initiative, along with the Army's accession mission and other prevailing policies. Nonetheless, this scenario represents the highest levels of recruiting and retention seen since 9/11.

Figure 5.2 shows that the Grow the Army results are in fact quite similar to those modeled under average conditions for the high-quality recruit mix. This likely reflects the fact that this period covered a range of economic conditions: civilian unemployment was low at the start of the period but shot up in FY 2008 during the financial crisis. In addition, our assumptions about enlistment bonuses, TV advertising, and SRBs are all largely based on the amounts the Army paid during the Grow the Army period.

We do note, however, that the Army expanded its enlistment eligibility criteria to some extent during the Grow the Army period. Why did this expansion not lead to the sort of rapid increase in accessions seen in Figure 5.3? Although the number of contracts written increased substantially (starting from an average of 90,000 per year between FYs 2006 and 2008 and reaching 102,000 in FY 2009), the accession mission remained at 80,000, as did the number of accessions.[1] These facts suggest that, had the Army wanted to access more lower-quality recruits during this time, it likely would have been able to do so.

Implications for NCO Experience

Regardless of external conditions, any regeneration scenario will result in a surge in new soldiers, requiring that more-senior enlisted soldiers be promoted more quickly to retain a fairly stable AC force shape. One potential concern is that such rapid promotion could lead to a substantial lowering of experience among midgrade and senior NCOs. As noted in Chapter Four, the inventory model allowed promotions only within the windows specified by the Army and required at least three YOS before promotion to E-5 and five YOS before promotion to E-6.

We measured average NCO experience by calculating the average YOS for soldiers in any given grade during each year of the expansion.[2] Table 5.1 shows the average YOS in steady state (before expansion) and the minimum average YOS during any of the expansion years. We focus here on three sets of scenarios: those requiring a high-quality recruit mix (lesser enlistment eligibility), those allowing a lower-quality recruit mix (greater enlistment eligibility), and the Grow the Army scenario. For the high-quality and lower-quality scenarios, we assumed that the Army has more recruiters and offers a higher SRB.

For all grades, average experience falls during the expansion. The effects are most pronounced for grades E-5 and E-6, in which the average number of YOS falls by more than 20 percent (from 6.6 years to around 5 years for E-5, and from 10.2 years to around 8 years for E-6). These numbers represent the *average* YOS for soldiers in these grades; to make room for the large cohorts of incoming soldiers, many soldiers are promoted to grade E-5 at three YOS and to grade E-6 at five YOS.

---

[1] Based on RA Analyst data.

[2] We took total man-YOS in a grade and divided by total number of soldiers in that grade.

Table 5.1
920K Scenario: Estimated NCO Experience

	E-5		E-6		E-7		E-8		E-9	
	Steady State	Minimum	Steady State	Minimum	Steady State	Minimum	Steady State	Minimum	Steady State	Minimum
Lesser enlistment eligibility (high-quality mix)										
Favorable conditions	6.6	5.2	10.2	7.8	16.2	14.7	19.1	17.7	24.1	22.9
Average conditions	6.6	5.3	10.2	7.8	16.2	14.7	19.1	17.7	24.1	22.9
Unfavorable conditions	6.6	5.3	10.2	7.9	16.2	14.7	19.1	17.7	24.1	22.9
Greater enlistment eligibility (lower-quality mix)										
Favorable conditions	6.6	4.3	10.2	8.5	16.2	14.7	19.1	17.3	24.1	22.9
Average conditions	6.6	4.5	10.2	8.5	16.2	14.1	19.1	17.3	24.1	22.9
Unfavorable conditions	6.6	4.8	10.2	8.0	16.2	14.9	19.1	17.2	24.1	22.9
Grow the Army	6.6	5.2	10.2	7.8	16.2	14.7	19.1	17.7	24.1	22.9

NOTES: *Average conditions* refers to an unemployment rate of 6.5 percent; *favorable conditions* refers to an unemployment rate of 8 percent; and *unfavorable conditions* refers to an unemployment rate of 5 percent. The Grow the Army scenario is based on recruiting and retention rates observed during FYs 2007–2009.

In Appendix B, we explore whether maintaining a "regeneration wedge"—that is, additional midgrade officers and NCOs who can serve as leaders in new units created under any expansion—can mitigate the decreases in average experience shown in Table 5.1. A regeneration wedge changes the shape of the force but not its overall starting or objective conditions. The appendix reproduces the results from Table 5.1 with a regeneration wedge. The results suggest that, with five years for regeneration, the resulting changes in average experience are fairly small, and the benefits associated with the changes would have to be compared against the cost of maintaining the wedge during peacetime.

Cost

Table 5.2 shows the incremental costs associated with increased accessions and retention—that is, how much activating these policies is likely to cost the Army to meet its growth objectives. For accessions, costs range from $776 million to $1.02 billion for the high-quality (lesser enlistment eligibility) scenarios. The costs are lower under unfavorable conditions because competing forces are at work. On one hand, the Army is competing with a strong civilian labor market, which would tend to increase bonus levels. On the other hand, the Army can find fewer recruits during unfavorable conditions, which would tend to decrease the overall payout. Recall that we capped per-recruit bonuses at their FY 2006–2008 levels; presumably, if the Army were willing to raise bonus levels without limit, it could meet its recruiting targets under virtually *any* conditions. Comparing the total amount of bonus payouts for the lower- and high-quality scenarios shows that, while the average *level* of bonuses paid is lower under

Table 5.2
920K Scenario: Estimated Incremental Costs of Recruiting and Retention

	Average Incremental Annual Costs	
	Accessions	Retention
Lesser enlistment eligibility (high-quality mix)		
Favorable conditions	1,021	475
Average conditions	1,001	473
Unfavorable conditions	776	472
Greater enlistment eligibility (lower-quality mix)		
Favorable conditions	1,144	474
Average conditions	1,247	475
Unfavorable conditions	1,222	477

NOTES: Numbers in millions of 2015 dollars. *Average conditions* refers to an unemployment rate of 6.5 percent; *favorable conditions* refers to an unemployment rate of 8 percent; and *unfavorable conditions* refers to an unemployment rate of 5 percent.

the lower-quality scenarios, the total *amount* is higher because of the large number of recruits.

The estimated incremental costs for increasing the use and size of SRBs to retain soldiers are roughly $475 million, with very little variation in expected incremental costs across the different scenarios. The costs vary little across scenarios because increasing SRBs induces very few additional soldiers to reenlist *who would not otherwise have reenlisted.* Recall that we estimated that increasing the SRB would increase the probability of reenlistment by only 3.5 percentage points (under the conservative elasticity estimate) and would apply only to those eligible for reenlistment who are near the expiration of their current terms of service. These small effect sizes mean that the variation across scenarios is vastly outweighed by the total incremental costs of introducing the program. We assume that bonuses would go from being a little-used tool for just a fraction of soldiers to near universal application. This means the program pays bonuses even to soldiers who would have reenlisted without it, and these costs dominate.

It is worth noting that, although the incremental costs of recruitment and retention bonuses are large, they are small relative to total compensation for the additional soldiers regeneration requires. Meeting targets from a base of 920K would require an additional 127,000 enlisted soldiers. In a 2007 report, the CBO estimated that average cash, noncash, and deferred cash compensation to a married, enlisted soldier with rank E-4 was approximately $89,700 (equal to approximately $105,500 in 2015 dollars).[3] A very rough estimate of the ballpark for the incremental annual compensation cost can be found by multiplying the number of additional soldiers (127,000) by average E-4 compensation in 2015 dollars. This exercise suggests that the incremental, annual compensation cost would be in the range of $13 billion.

Implications for the Reserve Component

What do the estimated shortfalls in AC size imply for the RC? We begin by considering the immediate demand scenario—in which the full target of deployable forces (with AC operating at 1:1 and RC at 1:4) is required immediately, as depicted in Figure 5.4.

Table 5.3 shows the estimated, *cumulative* MOB:Dwell ratio that the RC would need to achieve to backfill the AC shortfall, if the scenario required the target number of deployable forces immediately. In this case, since we assumed that no new AC recruits would be available for the first three years (one year of preparation and two years of recruiting and training), the full shortfall would need to be covered by the RC for the first three years, resulting in a MOB:Dwell ratio of 1:1.6.

The additional AC forces begin to deploy at the end of Year 3, with the final new surge cohort deploying at the end of Year 5. Thus the stress on the RC diminishes, so that the cumulative MOB:Dwell ratio over a six-year period would be approximately

---

[3]   Estimate of 2007 compensation is from CBO, 2007, Tab. 2. Dollar value is translated into 2015 dollars using the online inflation calculator at BLS, undated a.

Figure 5.4
Immediate Demand

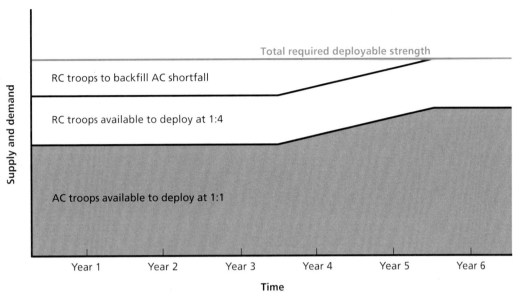

RAND RR1637-5.4

Table 5.3
920K Scenario: Estimated MOB:Dwell Ratio Assuming Immediate Increase in Demand

	Years 1–3	Year 4	Year 5	Year 6
Lesser enlistment eligibility (high-quality mix)				
Favorable conditions	1:1.6	1:1.7	1:1.8	1:2
Average conditions	1:1.6	1:1.7	1:1.8	1:1.9
Unfavorable conditions	1:1.6	1:1.7	1:1.7	1:1.9
Greater enlistment eligibility (lower-quality mix)				
Favorable conditions	1:1.6	1:1.7	1:1.9	1:2.1
Average conditions	1:1.6	1:1.7	1:1.9	1:2.1
Unfavorable conditions	1:1.6	1:1.7	1:1.9	1:2.1
Grow the Army	1:1.6	1:1.7	1:1.8	1:1.9

NOTES: *Average conditions* refers to an unemployment rate of 6.5 percent; *favorable conditions* refers to an unemployment rate of 8 percent; and *unfavorable conditions* refers to an unemployment rate of 5 percent. The Grow the Army scenario is based on recruiting and retention rates observed during FYs 2007–2009.

1:2. This does not mean that the RC would still be deployed at 1:2 at the end of Year 6 but, rather, reflects the average ratio of the preceding six-year period.

Table 5.3 shows that lowering recruit quality could help to ease the stress on the RC slightly in the later years; however, even in this case, the cumulative MOB:Dwell ratio would average 1:2.1 over the six-year period.

We also considered an alternative scenario in which demand begins at a base level equal to the deployable capacity of the existing forces (920K or 980K Army, with AC operating at 1:1 and RC at 1:4), then gradually builds up to the total target deployable strength (Figure 5.5). We assumed that demand ramps up linearly from the base level to the target level (so that one-quarter of the ultimate increase in troops is needed by the end of Year 1, one-half by Year 2, and so on).

Table 5.4 shows that, in this case, the MOB:Dwell ratio starts at 1:3.1 during Year 1, falls to 1:2.3 during Year 4 (by which time demand is fully ramped up, but the first cohort of AC surge troops has just deployed), then inches back up to around 1:2.4 by the end of Year 6 as more AC troops become available.

An alternative way to approach this issue would be to assume that the RC rotates at a certain speed, then estimate the shortfall in available troops using different external conditions and policies. Our baseline assumptions already factor in a 1:4 MOB:Dwell for the RC. Therefore, in the case of an immediate buildup in demand, the shortfalls would be identical to the shortfalls in AC capacity. We also tested the sensitivity of the results to assuming that the RC rotates at 1:3. In this case, the interim shortfalls would be lower, but would persist. Ultimately, our finding that the shortfall persists even at

Figure 5.5
Gradual Buildup from Base Level of Demand

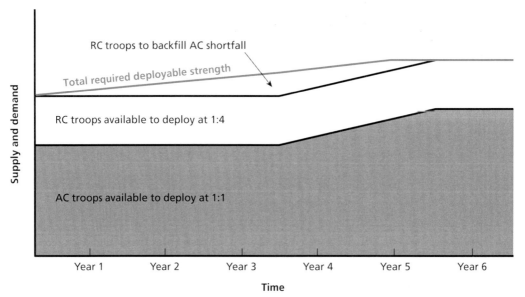

Table 5.4
920K Scenario: Estimated MOB:Dwell Ratio Assuming Linear Increase in Demand

	Year 1	Year 2	Year 3	Year 4	Year 5	Year 6
Lesser enlistment eligibility (high-quality mix)						
Favorable conditions	1:3.1	1:2.7	1:2.4	1:2.3	1:2.3	1:2.5
Average conditions	1:3.1	1:2.7	1:2.4	1:2.3	1:2.3	1:2.4
Unfavorable conditions	1:3.1	1:2.7	1:2.4	1:2.3	1:2.2	1:2.3
Greater enlistment eligibility (lower-quality mix)						
Favorable conditions	1:3.1	1:2.7	1:2.4	1:2.4	1:2.5	1:2.7
Average conditions	1:3.1	1:2.7	1:2.4	1:2.3	1:2.5	1:2.7
Unfavorable conditions	1:3.1	1:2.7	1:2.4	1:2.3	1:2.4	1:2.6
Grow the Army	1:3.1	1:2.7	1:2.4	1:2.3	1:2.3	1:2.4

NOTES: *Average conditions* refers to an unemployment rate of 6.5 percent; *favorable conditions* refers to an unemployment rate of 8 percent; and *unfavorable conditions* refers to an unemployment rate of 5 percent. The Grow the Army scenario is based on recruiting and retention rates observed during FYs 2007–2009.

the end of Year 5 (beginning of Year 6) unless enlistment eligibility standards are lowered holds, even if the RC rotates at 1:3.

980K Scenario

Effects of Accession and Retention Policies Under Average Conditions

When starting from a Total Army of 980K (AC of 450K), a combination of increased recruiters, TV advertising, enlistment bonuses, and SRBs can come close to achieving targets under average recruiting conditions, with a shortfall of only around 15,000 soldiers at the end of Year 5 (Figure 5.6).

Similar to the 920K scenario, lowering quality can close the shortfall completely by Year 5. However, in the 980K scenario, the shortfall is fairly small even while maintaining lesser enlistment eligibility (a high-quality mix). For the high-quality mix, the number of soldiers accessed per year in the 980K scenario is around 80,000 to 85,000, only slightly higher than in the 920K scenario.[4] However, because the required increase in the number of soldiers is substantially lower for the 980K scenario, the shortfall is also smaller.

---

[4] The number of accessions estimated in the 980K scenario is due to the higher number of recruiters associated with the larger starting force size.

Figure 5.6
980K Scenario: Estimated Enlisted Shortfall Using Different Policy Levers

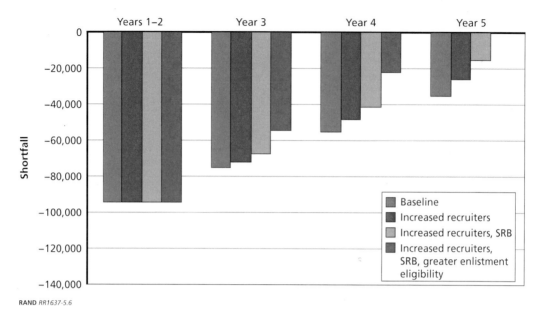

RAND RR1637-5.6

Figure 5.7
980K Scenario: Estimated Enlisted Shortfall Under Different Conditions

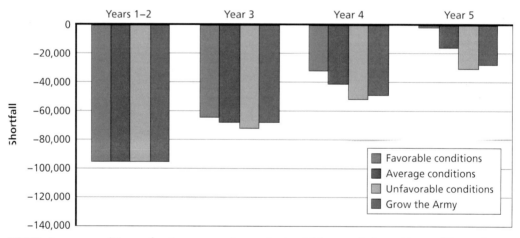

NOTES: *Average conditions* refers to an unemployment rate of 6.5 percent; *favorable conditions* refers to an unemployment rate of 8 percent; and *unfavorable conditions* refers to an unemployment rate of 5 percent. The Grow the Army scenario is based on 80,000 recruits per year and retention rates observed during FYs 2007–2009.

RAND RR1637-5.7

Effects of Different External Conditions

Figure 5.7 shows the effect of external conditions on the shortfall, assuming that the Army maintains lesser enlistment eligibility (high-quality mix) and uses enlistment bonuses, TV advertising, increased recruiters, and SRBs. Under favorable recruiting conditions, the shortfall is virtually eliminated by the end of Year 5. In contrast, if recruiting conditions are unfavorable, the shortfall can double to more than 30,000. If enlistment eligibility is expanded (quality is lowered), external conditions have little effect on the shortfall; the number of recruits is larger than can be accommodated under any conditions (Figure 5.8).

Figure 5.7 also shows the results from the Grow the Army scenario. In this case, the shortfall is approximately 28,000 at the end of Year 5, similar to the unfavorable conditions scenario. The reason the Grow the Army scenario more closely mirrors the unfavorable conditions scenario in this 980K case, even though it mirrored the average conditions scenario in the 920K case, is that it caps accessions at 80,000 per year. With a larger base force (980K rather than 920K), more accessions are needed simply to maintain the existing force; thus, *incremental* accessions are effectively lower.

Table 5.5 shows that, as in the 920K scenario, the 980K scenario would also entail a reduction in average YOS for NCOs. The size of the effect is approximately the same for E-5 because promotions to E-5 must still occur rapidly to accommodate a large influx of soldiers. However, the reduction in average YOS is less stark for E-6, falling from 10.2 YOS to 8.5 YOS under average conditions with lesser enlistment eligibility (albeit still representing a fall of almost 20 percent). Appendix B presents corresponding results for the 980K scenario with a regeneration wedge. As in the 920K scenario, the results show that a wedge does not make a substantial difference in terms of average experience, given the five-year time line.

Figure 5.8
980K Scenario: Estimated Enlisted Shortfall Under Different Conditions, Greater Enlistment Eligibility

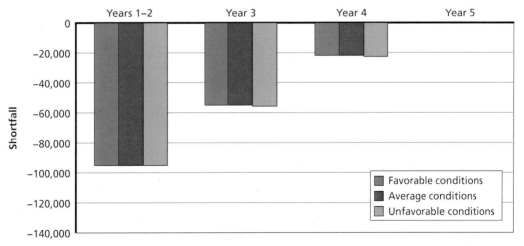

NOTES: *Average conditions* refers to an unemployment rate of 6.5 percent; *favorable conditions* refers to an unemployment rate of 8 percent; and *unfavorable conditions* refers to an unemployment rate of 5 percent.

RAND *RR1637-5.8*

Table 5.5
980K Scenario: Estimated NCO Experience

	E-5		E-6		E-7		E-8		E-9	
	Steady State	Minimum	Steady State	Minimum	Steady State	Min mum	Steady State	Minimum	Steady State	Minimum
Lesser enlistment eligibility (high-quality mix)										
Favorable conditions	6.6	5.2	10.2	8.5	16.2	14.8	19.1	17.7	24.0	23.1
Average conditions	6.6	5.2	10.2	8.5	16.2	15.2	19.1	17.8	24.0	23.1
Unfavorable conditions	6.6	5.3	10.2	8.5	16.2	15.2	19.1	17.9	24.0	23.1
Greater enlistment eligibility (lower-quality mix)										
Favorable conditions	6.6	4.5	10.2	9.2	16.2	14.2	19.1	17.5	24.0	23.5
Average conditions	6.6	4.5	10.2	9.2	16.2	14.2	19.1	17.5	24.0	23.5
Unfavorable conditions	6.6	4.7	10.2	9.0	16.2	14.4	19.1	17.4	24.0	23.1
Grow the Army	6.6	5.2	10.2	8.6	16.2	15.2	19.1	17.9	24.0	23.1

NOTES: *Average conditions* refers to an unemployment rate of 6.5 percent; *favorable conditions* refers to an unemployment rate of 8 percent; and *unfavorable conditions* refers to an unemployment rate of 5 percent. The Grow the Army scenario is based on 80,000 recruits per year and retention rates observed during FYs 2007–2009.

Table 5.6
980K Scenario: Estimated Incremental Costs of Recruiting and Retention

	Average Incremental Annual Costs	
	Accessions	Retention
Lesser enlistment eligibility (high-quality mix)		
Favorable conditions	1,038	500
Average conditions	988	501
Unfavorable conditions	746	500
Greater enlistment eligibility (lower-quality mix)		
Favorable conditions	1,084	495
Average conditions	1,210	495
Unfavorable conditions	1,194	495

NOTES: Numbers in millions of 2015 dollars. *Average conditions* refers to an unemployment rate of 6.5 percent; *favorable conditions* refers to an unemployment rate of 8 percent; and *unfavorable conditions* refers to an unemployment rate of 5 percent.

Cost

Table 5.6 shows the incremental costs associated with increased accessions and retention. For accessions, costs range from $746 million to $1.038 billion for the high-quality scenarios. As with the 920K scenario, costs are higher for greater enlistment eligibility (lower-quality mix) scenarios because of the larger number of recruits. The estimated incremental costs for retention under the 980K scenario are roughly $500 million, again with very little variation in expected incremental costs across the different recruiting environments.

Implications for the Reserve Component

Table 5.7 summarizes cumulative MOB:Dwell rates for the RC given immediate demand for the target number of deployable forces. Under the immediate demand scenario, the RC would need to rotate at 1:2 during the first 3 years, with the cumulative ratio decreasing to about 1:2.4 by Year 6. Building the AC faster by lowering recruit quality reduces the cumulative stress on the RC slightly by Year 6.

A more gradual demand buildup (Table 5.8) requires a MOB:Dwell ratio of around 1:3 during the first two years, rising to 1:2.7 by Year 4, by which time the full complement of deployable troops is required, but when only the first AC surge troops are ready to deploy. The ratio then tapers off slightly by the end of Year 6.

As in the 920K scenario, we can estimate the shortfall in available troops under the assumption that the RC rotates at 1:3. Recall that, with an RC rotation of 1:4, the shortfall could be eliminated at the end of Year 5 under favorable recruiting conditions, even if eligibility standards were not lowered. With an RC rotation of 1:3, the shortfall can be eliminated at the end of Year 5, even under *average* recruiting conditions.

Table 5.7
980K Scenario: Estimated MOB:Dwell Ratio Assuming Immediate Increase in Demand

	Years 1–3	Year 4	Year 5	Year 6
Lesser enlistment eligibility (high-quality mix)				
Favorable conditions	1:2	1:2.1	1:2.3	1:2.5
Average conditions	1:2	1:2.1	1:2.2	1:2.4
Unfavorable conditions	1:2	1:2.1	1:2.2	1:2.3
Greater enlistment eligibility (lower-quality mix)				
Favorable conditions	1:2	1:2.2	1:2.3	1:2.5
Average conditions	1:2	1:2.2	1:2.3	1:2.5
Unfavorable conditions	1:2	1:2.2	1:2.3	1:2.5
Grow the Army	1:2	1:2.1	1:2.2	1:2.4

NOTES: *Average conditions* refers to an unemployment rate of 6.5 percent; *favorable conditions* refers to an unemployment rate of 8 percent; and *unfavorable conditions* refers to an unemployment rate of 5 percent. The Grow the Army scenario is based on 80,000 recruits per year and retention rates observed during FYs 2007–2009.

Table 5.8
980K Scenario: Estimated MOB:Dwell Ratio Assuming Linear Increase in Demand

	Year 1	Year 2	Year 3	Year 4	Year 5	Year 6
Lesser enlistment eligibility (high-quality mix)						
Favorable conditions	1:3.3	1:3	1:2.8	1:2.7	1:2.8	1:2.9
Average conditions	1:3.3	1:3	1:2.8	1:2.7	1:2.7	1:2.8
Unfavorable conditions	1:3.3	1:3	1:2.8	1:2.7	1:2.7	1:2.7
Greater enlistment eligibility (lower-quality mix)						
Favorable conditions	1:3.3	1:3	1:2.8	1:2.7	1:2.8	1:3
Average conditions	1:3.3	1:3	1:2.8	1:2.7	1:2.8	1:3
Unfavorable conditions	1:3.3	1:3	1:2.8	1:2.7	1:2.8	1:3
Grow the Army	1:3.3	1:3	1:2.8	1:2.7	1:2.7	1:2.8

NOTES: *Average conditions* refers to an unemployment rate of 6.5 percent; *favorable conditions* refers to an unemployment rate of 8 percent; and *unfavorable conditions* refers to an unemployment rate of 5 percent. The Grow the Army scenario is based on 80,000 recruits per year and retention rates observed during FYs 2007–2009.

CHAPTER SIX

Conclusions and Implications for Preparation

In this chapter, we discuss some key findings from the analysis of the Grow the Army effort and discuss the modeling results. We also recommend specific measures the Army should take to prepare for future regeneration. Recall that this analysis applies to only one general type of contingency: a long-term, large-scale counterinsurgency and stability operation. For short-notice, intense operations, such as potential contingencies in the Baltic States or in Korea, the Army probably cannot achieve a useful degree of expansion in time to meet sudden demands for additional combat power.

The sum of our analyses indicates that it would probably be feasible for the Army to regenerate its deployable AC end strength within five years, with the first meaningful increase in Regular Army operational capacity becoming available at roughly the start of the third year. In this context, regeneration connotes expanding the Regular Army from a total size of 980K (AC of 450K) or 920K (AC of 420K)—the two future end strengths considered in the 2014 Army Posture Statement (McHugh and Odierno, 2014)—to meet the demands of a protracted, large-scale counterinsurgency or stability operation or operations of approximately the same aggregate scale of those in Iraq and Afghanistan at their peak.

However, while our analyses did not uncover any constraints that would make such regeneration infeasible, they suggest that the effort would carry a number of risks, particularly when expanding from a Total Army of 920K. What may be the most critical risk relates to the fact that, while the Regular Army is expanding, the Army as a whole will still need to meet operational demands. Thus, the Army will have to draw on its RC to an unprecedented extent to sustain high levels of operational commitment until it accomplishes regeneration. The Army will also need to be able to leverage extensive contract support throughout the duration of the conflict. As we discuss in more detail later in this chapter, the required rotation of the RC that we estimated should be feasible under the current authorities. However, RC forces may require mobilization periods exceeding the DoD's current one-year limit for involuntary mobilization to enable the forces to achieve the standard of proficiency needed to replace Regular Army forces for a useful period at acceptable risk. In addition, although ARNG leadership has expressed a willingness to operate at a tempo of 1:2, post-9/11 experience suggests that doing so may erode congressional or public support for sustained use of the RC.

Successful regeneration also depends on key defense leaders making tough decisions early, in advance of a clear demand signal for a larger Army. These include the decision to expand the Army itself, as well as supporting decisions to relax eligibility criteria, increase incentives, and offer the incentives to a wider range of individuals. In short, successful regeneration depends on spending a considerable amount of money and manpower as a hedge against things going wrong. Even if the necessary decisions are made, there is no guarantee that the Army will be able to implement them fully. Many different analyses have shown that the Army will struggle to obtain the necessary manpower if the economy is strong and unemployment is low.

The following sections expand on these broad findings and recommend measures to mitigate the risks identified in the course of this analysis. Besides making important decisions early, key recommendations include the needs to identify a specific contingency or contingencies that would require regeneration, to maintain adequate infrastructure and training capacity to accommodate a large surge in Army accessions, and to prepare the RC to support frequent and extensive mobilizations.

Major Findings

Current Policy Levers Will Probably Suffice to Enable Regeneration

Our research indicates that the current suite of policy levers will probably suffice to enable the Army to regenerate at the scale and speed desired within the overall context of the All Volunteer Force. Table 6.1 summarizes the policy options available for regenerating the Army. It is important to note that many of these policy options—notably, increasing end strength and various options for mobilizing the RC—rely on decision-makers outside the Army.

When the Army last grew, in response to post-9/11 demand, senior leadership used many of the options available (as indicated in Table 6.1).[1] These options were used at different points over the growth period and to varying degrees. End strength increases, both temporary and permanent, enabled the Army to grow by increasing the ceiling on the authorized number of soldiers across all three components. Force management options, such as adjusting BOG:Dwell and MOB:Dwell ratios, helped the Army increase the number of deployable troops. Contractor support enabled the Army to meet demand without further growing or stressing its forces. In the realm of recruiting and retention, the Army increased its capacity through a variety of financial incentives, waiver and eligibility changes, and involuntary stop-loss policies.

Our modeling took increases in end strength as givens and focused on the recruiting, retention, and force-management options that could be used to meet regeneration

---

[1]   Note that, depending on the circumstances, Individual Ready Reserve call-up can take place under presidential reserve call-up (10 USC 12304) authority or under partial mobilization (10 USC 12302) authority.

Table 6.1
Policy Options to Increase Army Capacity

End Strength Increases	Force Management	Recruiting and Retention	RC Mobilization Authorities
Exceed end strength caps[a]	BOG:Dwell[a]	Draft[b]	Full mobilization—10 USC 12301(a)
Permanent end strength increases[a]	MOB:Dwell[a]	Recruiting incentives[a]	Recall of retired reservists under full mobilization—10 USC 12301(a)
Temporary end strength increases[a]	Civilian support[a]	Retention incentives[a]	15-day statute—10 USC 12301(b)
	Contractor support[a]	Stop loss[a]	ADOS[a]—10 USC 12301(d)
		Waivers[a]	Partial mobilization[a]—10 USC 12302
			Presidential reserve call-up—10 USC 12304
			Reserve emergency call-up or disaster response activation—10 USC 12304a
			Activation for preplanned missions in support of combatant commanders—10 USC 12304b

[a] Observed during post-9/11 buildup.
[b] Selective Service System.

targets. Table 6.2 summarizes some of the key model results. Starting from a Total Army of 920K would likely result in a shortfall at the end of Year 5 if a high-quality recruit mix is maintained (i.e., enlistment eligibility is not expanded). When starting from 980K, it may be possible to eliminate the shortfall by the end of Year 5 while maintaining a high-quality recruit mix but only if recruiting conditions are favorable. In both cases, however, our modeling suggests that, if the Army is willing to combine generous incentives and aggressive advertising with expanded eligibility criteria, it may be possible to recruit the manpower needed to both sustain and expand the force.

These analyses implicitly assume an operational environment as permissive as that in Iraq and Afghanistan, in which the Army should be able to contract for many sustainment and support services. Similarly, the Army would need to generate a substantial portion of required operating forces using the RC. As discussed in Chapter Four, the first deployable surge AC troops will not be available until the end of Year 3, and the final surge AC troops will be deployed at the end of Year 5. Therefore, in all the cases we considered, the RC will be called on to rotate at a tempo of less than 1:3 over a six-year period. We discuss the feasibility of this rotation tempo in more detail later.

Table 6.2
Summary of Key Modeling Results

Scenario	Recruiters	Quality Mix	Recruiting Conditions	Average Annual Accessions	Shortfall at End of Year 5	Cumulative RC MOB:Dwell (immediate demand) at		Cumulative RC MOB:Dwell (gradual buildup in demand) at	
						Year 3	Year 6	Year 3	Year 6
920K	High	High	Favorable	87,051	25,578	1:1.6	1:2	1:2.4	1:2.5
920K	High	High	Average	81,088	39,975	1:1.6	1:1.9	1:2.4	1:2.4
920K	High	High	Unfavorable	74,990	54,620	1:1.6	1:1.9	1:2.4	1:2.3
920K	High	Low	Favorable	104,552	—	1:1.6	1:2.1	1:2.4	1:2.7
920K	High	Low	Average	103,526	—	1:1.6	1:2.1	1:2.4	1:2.7
920K	High	Low	Unfavorable	99,626	—	1:1.6	1:2.1	1:2.4	1:2.6
920K	Grow the Army	Grow the Army	Grow the Army	80,000	44,552	1:1.6	1:1.9	1:2.4	1:2.4
980K	High	High	Favorable	89,985	1,334	1:2	1:2.5	1:2.8	1:2.9
980K	High	High	Average	83,428	15,236	1:2	1:2.4	1:2.8	1:2.8
980K	High	High	Unfavorable	77,141	30,688	1:2	1:2.3	1:2.8	1:2.7
980K	High	Low	Favorable	97,800	—	1:2	1:2.5	1:2.8	1:3
980K	High	Low	Average	97,800	—	1:2	1:2.5	1:2.8	1:3
980K	High	Low	Unfavorable	97,326	—	1:2	1:2.5	1:2.8	1:3
980K	Grow the Army	Grow the Army	Grow the Army	80,000	27,712	1:2	1:2.4	1:2.8	1:2.8

Conditions Matter

Even though existing policy levers may suffice to support a large-scale regeneration, there is no definitive answer to whether regeneration will be feasible under specific future circumstances. Feasibility depends both on external conditions and on the willingness of the Army, DoD, the President, and Congress to use existing policy tools and on the willingness of the American public to respond to their use. We modeled the effect that economic conditions might have on the Army's ability to recruit new soldiers; all other things being equal, as Table 6.2 illustrates, a low unemployment rate can increase the likely shortfall by tens of thousands of soldiers. But equally important are various external conditions that are not as easily quantified. Political conditions may affect congressional willingness to increase end strength caps rapidly. As we discuss in more detail later, rotating the RC at a MOB:Dwell ratio faster than 1:3 may not be feasible, given the political climate.

Specific conditions of the conflict will also matter. During the Grow the Army period, contractors were extensively employed in providing training, procurement, and other support, both domestically and in the conflict areas. If the future conflict zone is not considered sufficiently safe or appropriate for contractors, the demands on the military may be much higher, requiring an even larger increase in troop size. In addition, during the OEF/OIF conflicts, a number of Army positions were filled by soldiers from other services (see, for example, Bates, 2007). To the extent that this may not be possible in future conflicts, future regeneration requirements for the Army may be higher than those we have assumed. The pace of the increase in demand for deployable troops is also a critical factor, particularly in determining how quickly the RC will need to rotate while building up the AC.

Feasibility will also depend on what policy levers the Army is willing to use. Our analysis assumed that the Army would employ enlistment bonuses and SRBs in the range of those seen during the Grow the Army initiative. If Congress authorizes and the Army offers substantially higher bonuses, then both accessions and retention could be much higher than projected, and regeneration could be more easily accomplished. It is also possible, though, that budgetary conditions will be less favorable than during Grow the Army and that the bonuses that the Army can offer will be less than those we considered in our analysis.

Regeneration Would Place Substantial Stress on the Reserve Component, Especially When Starting from 920K

All the regeneration scenarios we considered would require the RC to rotate at less than 1:3 for a number of years (Table 6.2). Table 6.3 summarizes the current set of authorities that govern the circumstances under which reservists can be activated and any restrictions on their use.

The post-9/11 conflicts relied on only two of these authorities: partial mobilization and active duty for operational support (ADOS). Partial mobilization was

Table 6.3
Accessing the RC for Operations

	Statutory Source	Authority	Utilization Process	Intended Use	Limitations
Involuntary	10 USC 12301(a) (full mobilization)	Congress	Requires congressional declaration of war or national emergency	Rapid expansion of armed forces to meet an external threat to national security	• No personnel limitation • Duration plus 6 months • Applicable to all reservists (inactive and retired)
	10 USC 12302 (partial mobilization)	President	Requires presidential declaration of national emergency (President must renew annually)	Manpower required to meet external threat to national emergency or domestic emergency	• Maximum 1 million Ready Reservists on active duty • Not more than 24 consecutive months • Used for OIF/OEF contingency operations
	10 USC 12304 (presidential reserve call-up for situations other than during war or national emergency)	President	Requires presidential notification to Congress	Augment AC for operational missions or support for domestic response to weapons of mass destruction or terrorist attacks	• Maximum 200,000 Ready Reservists on active duty • Maximum 30,000 Individual Ready Reservists • Limited to 365 days active duty • Prohibited for support of federal or state government during man-made or natural disasters
	10 USC 12304a (Assistance in response to a major disaster or emergency—Title 10 reserves only)	Secretary of Defense	Governor of affected state requests assistance Secretary of Defense may involuntarily order any unit to active duty following governor's request	Emergency response to national emergency and disasters	• No personnel limitation • Limited to continuous period of not more than 120 days
	10 USC 12301(b) (15-day statute)	Service secretary	Authority to order reservist to active duty without member's consent	Annual training or operational mission	• 15 days active duty once per year • Governor's consent required for National Guard

Table 6.3—Continued

	Statutory Source	Authority	Utilization Process	Intended Use	Limitations
	10 USC 12304b (preplanned missions in support of combatant commanders)	Service secretary	May order any unit to active duty to augment the AC Must submit to Congress a report on circumstances of order to active duty and follow prescribed policies and procedures	Augment AC for missions in support of combatant commander requirements	• Maximum of 60,000 on active duty at any one time • Limited to 365 consecutive days • Manpower and costs are specifically included and identified in the defense budget for anticipated demand • Budget information includes description of the mission and the anticipated length of time for involuntary order to AC
Voluntary	10 USC 12301(d) (ADOS)	Service secretary	Authority to order reservist to active duty with member's consent In case of National Guard, governor also must consent to voluntary activation	Operational missions (volunteers)	• Applicable to Ready Reserve • No duration • Governor's consent required for National Guard

SOURCES: RAND Arroyo Center compilation, October 2014.

also the most commonly used mobilization authority in previous conflicts. With the exception of ADOS (which is voluntary), the RC mobilization authorities available to senior leadership are dependent on the nature of the underlying conflict and the statutory authority granted to the President, Congress, or the service secretaries. For example, full mobilization enables the rapid expansion of the armed forces with all RC military personnel for the duration of the conflict plus six months, but exercising this authority requires a congressional declaration of national emergency or war. Both partial mobilization and presidential reserve call-up authority require presidential action before RC soldiers can be mobilized.[2] Mobilizing the RC under 10 USC 12304a in response to a national emergency or disaster requires both the affected governor's request for assistance and the Secretary of Defense's approval for mobilization of Title 10 reserves. Only mobilization under 10 USC 12301(b) (the 15-day statute), under 10 USC 12304b (preplanned missions in support of combatant commanders), or under 10 USC 12301(d) (voluntary mobilization under ADOS) can be exercised under the Secretary of the Army's authority alone.

Because mobilization authorities can be specific to certain types of conflict, the Army is limited in its use of this policy option to increase its capacity. For example, if there is an official declaration of a national emergency or war, reservists could be involuntarily mobilized under full mobilization or partial mobilization, the most expansive of the authorities. Conversely, if the next need is in response to a domestic terrorist attack or a natural disaster, the Army is limited to the mobilization authorities and their troop level and time restrictions specific to such events.

The required RC rotation that we estimated should be feasible under the current authorities. However, DoD policy calls for a maximum period of involuntary mobilization of one year (excluding individual skill training and postmobilization leave) (DoD Directive 1235.10, 2011). Some RC forces may require mobilization periods exceeding the current one-year policy limit to enable them to achieve the standard of proficiency needed to replace Regular Army forces for a useful period at acceptable risk.

DoD policy also aims for a 1:5 MOB:Dwell ratio for RC personnel, when possible (DoD Directive 1235.10, 2011). Although, as discussed in Chapter Two, ARNG leadership has expressed a willingness to operate at a tempo of 1:2 (Grass, 2013), post-9/11 experience suggests that doing so may erode congressional or public support for sustained use of the RC. One particular issue is that the current interpretation of partial mobilization authority—under which forces may be mobilized for no more than 24 consecutive months at a time but may be mobilized repeatedly—may be called into question.

---

[2] Partial mobilization requires a presidential declaration of national emergency that must be renewed annually (10 USC 12302). Presidential reserve call-up authority does not require a declaration of war or national emergency, but it does require written presidential notification to Congress (10 USC 12304(f)).

Recommendations

The following recommendations address actions that lie largely within the remits of DoD and the Army. These efforts require public support to succeed. Political leaders—especially the President—must be prepared to expend political capital to generate and sustain the necessary level of public support and create a context in which measures to expand the Army can succeed.

The Departments of Defense and Army Should Develop Planning Scenarios Requiring Regeneration

The study described here assessed the broad feasibility of regeneration at a certain scale and speed and identified the kinds of measures that the Army would need to accomplish that sort of regeneration. However, the Army needs to develop and resource specific capabilities to enable regeneration. Such capabilities include recruiters, institutional trainers, infrastructure equipment, and leaders in units, among other things. The Army must determine how many of these people or things it needs to maintain in the inventory and how many it must produce as part of regeneration. A conflict that required a more highly skilled mix of recruits or one that was fought where contractors could not be deployed would create much more serious challenges and would require additional preparation, such as retaining additional soldiers with certain skills or preparing to pay much higher enlistment bonuses. Developing and planning for specific scenarios would allow the Army to make more-informed decisions about how best to invest current resources in preparing for potential regeneration. More important, it would enable the Army to identify specific requirements, include them in its long-term program, and assess the degree to which the necessary capabilities are on hand.

Assess Alternative Ways to Posture the Army for Regeneration

To date, analysis has focused on how to generate the raw human capacity to staff an expanding Army. Considerably less thought has been devoted to posturing the Army to receive and manage the expansion. The Army has considered several different approaches to expansion over its history, including but by no means limited to: establishing cadre formations, undermanning units during peacetime to be filled out in war, drawing on manpower from its generating force, and relying on RC units to fill critical gaps in larger units (roundout). The Army should explore which of these approaches, or what combination thereof, best postures the Army to expand rapidly in time of crisis.

Prepare the Reserve Components for Rapid and High-Frequency Deployment

For the scenarios we considered, the RC will be called on at a rotational frequency at or below 1:3 over a six-year period under any external conditions. The chief of the National Guard Bureau has committed to supporting a BOG:Dwell of 1:2 (Grass, 2013), but the RC never reached that level in recent overseas contingency operations.

Our analysis indicates that the ARNG may have to make good on that commitment in the initial stages of a conflict. In fact, MOB:Dwell ratios in the vicinity of 1:3 raised political challenges in the early stages of the wars in Afghanistan and Iraq, and the deployment ratios that may be required for an immediate surge in demand, or when starting from 920K, could drive the ratio well below ratios seen in recent history. Moreover, the average ratio is likely to understate stress for RC units that are in high demand.

Thus, it will be critical to prepare the RC and its stakeholders today for the possibility of rapid, high-frequency deployment in the event of a conflict that requires regeneration. Our current research does not shed light on what particular form such preparation should take, but our findings make it clear that the Army will need to rely heavily on the RC for a substantial time. Failure to prepare stakeholders in advance for these commitments risks disruption in a time of crisis.

Preparation may also involve changing policies that limit RC employment. At the very least, it will be important to clarify the current interpretation of the partial mobilization authority—that is, mobilizing reserve forces for no more than 24 consecutive months at a time but allowing repeated mobilizations. In addition, current DoD policy limits mobilization (excluding individual skill training and postmobilization leave) to one year.[3] Shortages of Regular Army units may compel the Army to call on RC forces for more-demanding missions than they have been required to perform recently. As we noted in Chapter Two, RC brigades employed in such demanding roles in the 2005–2006 period conducted six months of predeployment training. Meeting the chief of the National Guard Bureau's commitment to provide forces on the ground for a whole year would certainly require exceeding the one-year limit established by current policy.

Maintain Certain Critical Skills in the Army Today to Reduce the Stress on the Army During Regeneration

We have focused on the Army's ability to bring in a sufficient number of "generic" recruits to rebuild the AC. However, it will not be so easy to fill a variety of specialized functions on short notice. For example, pilots, medical staff, and special operations forces typically require additional training, and it may be impossible to build such skills in a short time. Shortfalls in the AC for these specific functions may be mitigated to some extent by the use of soldiers from the RC or the Individual Ready Reserves who have the appropriate skills. The challenge is likely to be exacerbated by the fact that rapid regeneration—particularly starting from 920K—will likely require some lowering of the average quality of recruits, thus reducing the share of recruits from which positions in highly skilled MOSs can be filled.

A similar point of preparation today involves maintaining a wedge of additional midgrade officers and NCOs who can serve as leaders for the incoming surge of acces-

---

[3]   DoD Directive 1235.10, 2011.

sions during regeneration to enable rapid expansion of the Army in response to some future crisis. Our baseline analysis—which does not start with such a wedge—shows that regeneration is likely to decrease average YOS in grades E-5 and E-6 by 20 percent or more. Incorporating a wedge of soldiers in grades E-5 through E-9 mitigates the fall in average experience among these ranks but only by a small amount. Maintaining a wedge of leaders—as well as soldiers with MOSs that have long training lead times— has implications for both cost and readiness of today's force, which vary depending on where soldiers are stationed. Whether the costs associated with this approach are offset by the potential reduction in risk would thus be a fruitful avenue for future study.

Maintain Army Capacity for Contingency Contracting

As noted in Chapter Two, sustaining operations in Afghanistan and Iraq required in excess of 200,000 contractors during critical periods. In 2011, the Commission on Wartime Contracting concluded that contractors "have performed vital tasks in support of U.S. defense, diplomatic, and development objectives. But the cost has been high. Poor planning, management, and oversight of contracts has led to massive waste and has damaged these objectives" (Commission on Wartime Contracting in Iraq and Afghanistan, 2011, "Foreword"). The commission attributed these shortcomings to shortfalls in capability and capacity in the government's acquisition workforce. As the Army reduces its end strength, the natural tendency will be to reduce the acquisition workforce to levels commensurate with the supported force. The Army should resist that tendency. As Army operating forces decrease, the need to contract support and sustainment capacity may well increase and will certainly not decrease. In short, the ability to manage a large contracted workforce is likely to become even more critical to sustaining operational capacity.

Develop Contingency Plans

As we have already noted several times, our analysis has not identified any definitive limit to the Army's ability to regenerate at the speed and on the scale described in this analysis. It has, however, indicated that these efforts are fraught with risk. The maximum accessions the Army was able to achieve during the Grow the Army initiative were around 80,000 a year, even with expanded eligibility criteria. As discussed in the text, that was the Army's objective at the time, so we cannot assume it constitutes a limit. It may be a warning, however. For that reason, as the Army plans for rapid expansion of the Regular Army, it should also develop contingency plans should that expansion falter. Those contingency plans will almost certainly hinge on a much higher degree of mobilization of the Army's RC.

Decide Early

Our analysis shows that, given certain external conditions and assuming that the Army can use and is willing to use certain combinations of policy levers, meeting or approaching regeneration targets will be feasible. However, our analysis assumed that the decision to regenerate rapidly would be made at the start of the conflict and that all policy levers would be in place by the end of the first year of the conflict to deliver the first meaningful increment of capability by the third year of the conflict. For example, when regenerating from 920K, accepting a greater enlistment eligibility (lower-quality mix) of recruits will likely be necessary to meet targets by the end of the fifth year. We assumed that this decision would be made when recruiting is first ramped up. Expanding enlistment eligibility at the end of the third year, for example, would likely not be sufficient to make up for lower accessions in previous years and to meet targets by the end of the fifth year. Similarly, it takes time to increase the number of recruiters and train them and to increase advertising effort and have it pay off. We assumed that these policies would be put in place during the first year after the decision to expand has been made and that bonuses, TV advertising, and recruiter capacity are all being used at optimal levels by the end of the first year. If there is a concern that there will not be sufficient lead time to secure TV advertising and put recruiters into place, an alternative may be to maintain a higher entry DEP. However, previous research suggests that maintaining a higher entry DEP is likely to be cost effective only if there is not sufficient lead time to optimize TV advertising and incentives and only under favorable or average recruiting conditions (Orvis et al., 2016).

If decisions can be made even more quickly than we have assumed, increased recruiting and retention may even be possible during the first year. Similarly, we assumed that it takes two years from the time the Army starts trying to recruit additional soldiers for the new soldiers to complete both individual and unit training. To the extent that the training schedule can be compressed, the number of deployable troops can be increased (and the stress on the RC decreased) more quickly. An increased training pace would depend crucially on having even more trainers and training facilities available.

As noted in this chapter's introduction, these will be difficult decisions. In making them, officials will have to allocate significant amounts of scarce resources as a hedge against eventualities they wish to avoid. Committing resources to increased recruiting, retention, and training reduces the amount of resources—in terms of money and human capital—available for other, potentially equally important efforts. If events unfold according to plan, the Army will divest those regenerated capabilities without ever having used them. Doing so will expose officials to being called to account for "wasting" the resources. Nonetheless, it may be prudent, at the start of a new conflict, to explicitly prepare for a potential postconflict stabilization phase, which may require more capacity than the initial phase. Preparation need not necessarily increase the size of the onboard workforce but could involve such measures as requesting the authority to increase end strength if needed, buying options for increased TV advertising, and preidentifying onboard personnel who could serve as additional recruiters. If the experience of OIF and OEF has taught no other lesson, it is that events can seldom be counted on to go according to plan. As that experience has further indicated, having soldiers but not needing them incurs far less regret—albeit higher costs—than needing them and not having them. For that reason, a decision to go to war should be a decision to expand the Army.

Conclusion

This analysis indicates that regeneration is theoretically feasible, if the Army has prepared for that contingency and if DoD officials make and implement challenging and unpalatable decisions early. Our modeling demonstrates that expanding from both 980K and 920K is possible but that decisionmakers should be considerably more cautious about assuming that regeneration is a sufficient hedge against the 920K Army's potential capacity shortfalls.

Both regeneration scenarios also entail substantial risk. In all cases we considered, the RC forces would have to operate at MOB:Dwell rates significantly below 1:3, and in many cases below 1:2, for several years to sustain the Army's operational capacity. These ratios could decrease further if RC units are deployed for more-demanding missions and if their predeployment training thus takes longer. Meeting regeneration targets will also almost certainly require lowering eligibility standards immediately and keeping them low longer than they were during the Grow the Army effort.

Additional Modeling Results

This appendix presents the full suite of inventory modeling results. We begin by comparing the estimated shortfall under each of the five conditions examined in the retention model:

- a *baseline* case that takes average continuation rates from 2003–2012 from TAPDB data
- a *low* estimated effect from introducing reenlistment bonuses that increases these baseline continuation rates somewhat
- a corresponding *high* estimate from reenlistment bonuses that has a larger effect on these baseline continuation rates
- an *upper-bound* estimate that takes the average continuation rates drawn from FYs 2007–2009
- a corresponding *lower-bound* estimate that takes average continuation rates from FYs 2003–2005.

We focus here on the results that assume an increased number of recruiters and a high-quality accession mix. The baseline result is the same as that shown in Figure 5.1, with no SRB. The low-SRB effect result corresponds to the SRB result shown in Figure 5.2, while the high-SRB effect illustrates that the shortfall may be somewhat smaller with the more-optimistic reenlistment effects from the previous literature. Note that the more optimistic SRB effect yields a smaller shortfall than what we would expect if continuation rates were equal to the highest average rates seen in recent years (FYs 2007–2009, the upper-bound results); this may reflect the fact that FYs 2007–2009 included a period with a very tight civilian labor market. Nonetheless, the low-SRB, high-SRB, and upper-bound effects tell a similar story: that the shortfall would be in the range of 35,000 to 40,000 soldiers at the end of three years of surge recruiting when starting from 920K (Figure A.1).

The lower-bound results reflect continuation rates equal to the lowest seen in recent years (FYs 2003–2005). On one hand, these rates reflect a time when the Army was not trying to grow and may thus be too pessimistic to reflect a realistic regeneration scenario. On the other hand, these rates can be taken to illustrate the potential

Figure A.1
920K Scenario: Estimated Enlisted Shortfall Under Different Conditions

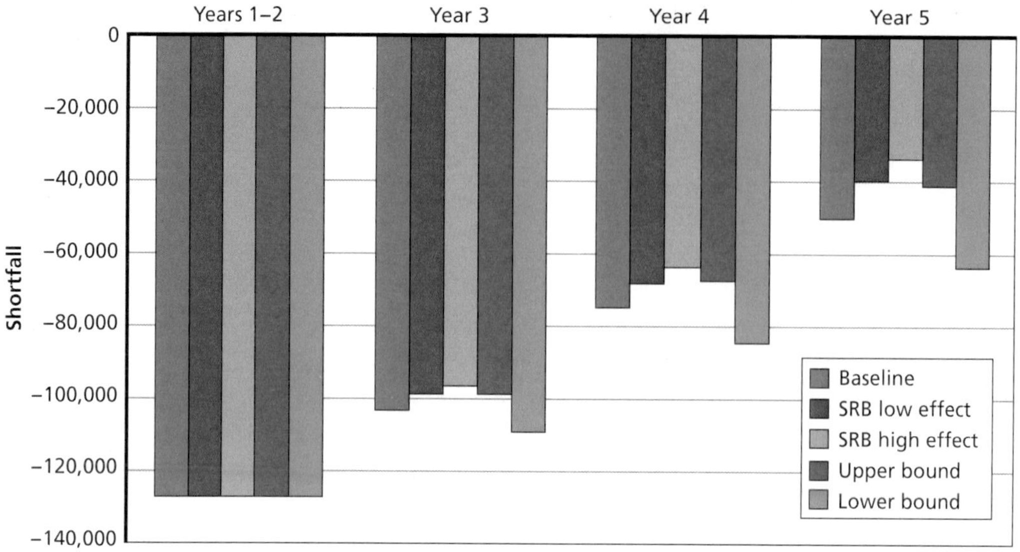

RAND *RR1637-A.1*

effects of decreased propensity to reenlist, even more-attractive labor market conditions than during FYs 2006–2007, or other factors that may lower continuation rates in the future. Such lower rates would result in a shortfall of more than 60,000 soldiers at the end of three years of surge recruiting when starting from 920K.

Figure A.2 shows results for the 980K scenario. The low-SRB, high-SRB, and upper-bound results all suggest a shortfall in the neighborhood of 10,000 to 15,000 soldiers at the end of three years of surge recruiting. The lower-bound continuation rates suggest a much higher shortfall of nearly 40,000, more in line with the shortfalls typically seen in the 920K scenario.

Tables A.1 and A.2 summarize shortfall estimates for the full range of scenarios considered.

Figure A.2
980K Scenario: Estimated EnlistedShortfall Under Different Conditions

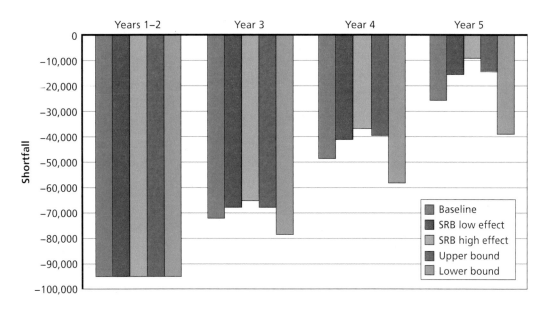

Table A.1
920K Scenario: Summary of All Results

Recruiters[a]	Enlistment Eligibility[b]	Recruiting Conditions[c]	Average Annual Accessions[d]	Shortfall, Year 5				
				Baseline	SRB Low Effect	SRB High Effect	Upper Bound	Lower Bound
Low	Lesser	Favorable	81,550	49,228	39,110	32,777	40,203	62,756
Low	Lesser	Average	77,091	59,733	49,632	43,408	50,935	73,139
Low	Lesser	Unfavorable	70,848	74,620	64,668	58,856	66,194	87,604
Low	Greater	Favorable	103,769	—	—	—	—	1,138
Low	Greater	Average	102,254	397	—	—	—	14,925
Low	Greater	Unfavorable	94,635	18,374	7,743	1,429	8,880	32,464
High	Lesser	Favorable	87,051	36,003	25,578	19,170	26,838	49,799
High	Lesser	Average	81,088	50,168	39,975	33,642	41,252	63,762
High	Lesser	Unfavorable	74,990	64,721	54,620	48,470	55,959	77,979
High	Greater	Favorable	104,552	—	—	—	—	—
High	Greater	Average	103,526	—	—	—	—	4,003
High	Greater	Unfavorable	99,626	6,597	—	—	—	20,994
Grow the Army[e]	Grow the Army[e]	Grow the Army[e]	80,000	54,573	44,552	38,405	45,187	67,729

[a] *Low* implies 5,433 OPRA foxhole recruiters; *high* implies 6,011 OPRA foxhole recruiters.

[b] *Lesser* implies that 55 percent of recruits are high quality, that there are no prior-service accessions, and that 10 percent of recruits receive waivers; *greater* implies that 45 percent of recruits are high quality, that there are 10,000 prior-service accessions, and that 20 percent of recruits receive waivers.

[c] *Favorable, average,* and *unfavorable* recruiting conditions correspond to unemployment rates of 8, 6.5, and 5 percent, respectively.

[d] This is the average accessions expected during the three surge recruiting years; this average includes only accessions that are actually used in the retention model, given force shaping constraints.

[e] *Grow the Army* corresponds to a scenario with 80,000 accessions per year.

Table A.2
980K Scenario: Summary of All Results

Recruiters[a]	Enlistment Eligibility[b]	Recruiting Conditions[c]	Average Annual Accessions[d]	Shortfall, Year 5				
				Baseline	SRB Low Effect	SRB High Effect	Upper Bound	Lower Bound
Low	Lesser	Favorable	85,669	20,232	9,606	4,590	8,897	33,837
Low	Lesser	Average	79,397	35,467	25,068	18,480	23,963	47,898
Low	Lesser	Unfavorable	73,000	50,925	40,785	34,312	39,777	62,642
Low	Greater	Favorable	97,800	—	—	—	—	—
Low	Greater	Average	97,800	—	—	—	—	—
Low	Greater	Unfavorable	96,544	—	—	—	—	6,158
High	Lesser	Favorable	89,985	10,567	1,334	—	799	24,435
High	Lesser	Average	83,428	25,818	15,236	9,178	14,152	38,788
High	Lesser	Unfavorable	77,141	40,991	30,688	24,121	29,604	53,057
High	Greater	Favorable	97,800	—	—	—	—	—
High	Greater	Average	97,800	—	—	—	—	—
High	Greater	Unfavorable	97,326	—	—	—	—	—
Grow the Army[e]	Grow the Army[e]	Grow the Army[e]	80,000	38,267	27,712	20,987	26,009	50,263

[a] *Low* implies 5,821 OPRA foxhole recruiters; *high* implies 6,440 OPRA foxhole recruiters.

[b] *Lesser* implies that 55 percent of recruits are high quality, that there are no prior-service accessions, and that 10 percent of recruits receive waivers; *greater* implies that 45 percent of recruits are high quality, that there are 10,000 prior-service accessions, and that 20 percent of recruits receive waivers.

[c] *Favorable, average,* and *unfavorable* recruiting conditions correspond to unemployment rates of 8, 6.5, and 5 percent, respectively.

[d] This is the average accessions expected during the three "surge" recruiting years; this average includes only accessions that are actually used in the retention model, given force shaping constraints.

[e] *Grow the Army* corresponds to a scenario with 80,000 accessions per year.

Sensitivity of Results with Regeneration Wedges

Tables 5.1 and 5.5 showed that average YOS declined under virtually all regeneration scenarios, particularly at lower ranks, as soldiers are quickly promoted through the ranks to fill expanding requirements for NCOs. One possible method for improving average YOS during rapid regeneration is to retain a regeneration wedge of midgrade officers and NCOs during a drawdown period who can serve as leaders in new units created under any expansion. We experimented with such a regeneration wedge as a robustness check to see whether and how it would affect our overall results and, in particular, how it would affect average YOS for enlisted soldiers. We considered the effects of this wedge for both the 920K and the 980K scenarios.

The regeneration wedge we tested made assumptions different from those in Chapter Five about the force *shape* but did not change the overall starting or objective conditions for total force *strength*. Under the wedge, authorizations are reallocated from lower enlisted ranks (E-1–E-4) to higher enlisted ranks (E-5–E-9) and to officer ranks (O-3–O-6). First, to increase the number of officers, we relaxed the assumption of enlisted soldiers constituting 80 percent of AC end strength and assumed instead that they constituted approximately 78.5 percent at the start, meaning a larger share of the total starting AC force were officers in Year 1. Doing so reduced starting enlisted authorizations from 360,500 under the 980K scenario to 355,500, and from 336,500 under the 920K scenario to 330,000. (By the end of the regeneration window, we assumed that enlisted soldiers once again constituted 80 percent of the total AC force.) This skimming of enlisted authorizations to provide additional officer authorizations was equally distributed (in terms of percentage reduction) across enlisted ranks. Additional officers would need to be retained and promoted over time to fill these authorizations. Second, we reallocated authorizations within the enlisted ranks, from E-1–E-4 to E-5–E-9. Slots were reallocated away from grades E-1 to E-4, and toward grades E-5 to E-9 in proportion to the original share of authorizations in each grade level.

Table B.1 summarizes the starting conditions for the regeneration wedge and how that compares with baseline. Note that the total number of enlisted authorizations is lower because some authorizations are moved over to the officer ranks.

We reran the various scenarios described in Chapter Five of the main report using this wedge to explore how it affected average experience levels and to determine the

Table B.1
Regeneration Wedge Starting Conditions

	980K Scenario			920K Scenario		
Grade	Baseline	Alternative Manning Profile	Wedge	Baseline	Alternative Manning Profile	Wedge
E9	3,384	3,647	263	3,158	3,502	344
E8	10,734	11,571	837	10,018	11,111	1,093
E7	37,829	40,777	2,948	35,307	39,159	3,852
E6	59,026	63,625	4,599	55,091	61,100	6,009
E5	73,123	78,820	5,697	68,248	75,693	7,445
E4	107,386	95,636	−11,750	100,228	84,975	−15,253
E1–E3	68,986	61,437	−7,549	64,387	54,588	−9,799
Total	360,468	355,513	−4,955	336,437	330,128	−6,309

resulting shortfalls of soldiers under a number of different external conditions. As in Chapter Five, we assumed the Army drew on all its available accession and retention policy levers—an increased level of recruiters, TV advertising, enlistment and reenlistment bonuses, and possibly relaxing eligibility criteria.

Shortfall results with the wedge were broadly similar to those in Chapter Five. We do not describe those results in detail here but rather focus on how the wedge affected average experience levels of the enlisted ranks. Recall that Tables 5.1 and 5.5 presented results on average experience levels of NCOs without any regeneration wedge for the 920K and 980K scenarios, respectively. Both Tables 5.1 and 5.5 showed that average experience levels fell for all grades, particularly E-5 and E-6, as soldiers were rapidly promoted to make room for incoming cohorts.

Table B.2 repeats the presentation of Table 5.1 from the main report, applying the results from the wedge to the 920K scenario to show the average YOS in steady state (before expansion) and the minimum average YOS during any of the three expansion years. As in Table 5.1, we focus on three sets of scenarios: those requiring a high-quality recruit mix (lesser enlistment eligibility), those allowing a lower-quality recruit mix (greater enlistment eligibility), and the Grow the Army scenario. For the high-quality and lower-quality scenarios, we assumed that the Army had increased recruiting resources optimally and offered a higher SRB.

Here, the average experience of soldiers at the lowest grade, E-5, is virtually identical to that found without a wedge (Table 5.1) when a high-quality recruitment mix is enforced. If more low-quality recruits are recruited, average experience levels slightly improve by roughly 0.3 years, on average, relative to the no-wedge option in Table 5.1. Larger improvements in average YOS are seen at grade E-6, which gains an average of

Table B.2
920K Scenario with Regeneration Wedge: Estimated NCO Experience

	E-5		E-6		E-7		E-8		E-9	
	Steady State	Minimum	Steady State	Minimum	Steady State	Minimum	Steady State	Minimum	Steady State	Minimum
Lesser enlistment eligibility (high-quality mix)										
Average conditions	6.6	5.3	10.2	8.5	16.2	15.3	19.1	18.0	24.0	23.2
Favorable conditions	6.6	5.3	10.2	8.5	16.2	15.4	19.1	17.7	24.0	23.2
Unfavorable conditions	6.6	5.5	10.2	8.5	16.2	15.3	19.1	18.0	24.0	23.2
Greater enlistment eligibility (lower-quality mix)										
Average conditions	6.6	4.8	10.2	9.2	16.2	14.4	19.1	17.8	24.0	23.2
Favorable conditions	6.6	4.6	10.2	9.2	16.2	14.4	19.1	17.8	24.0	23.2
Unfavorable conditions	6.6	5.2	10.2	8.5	16.2	15.2	19.1	17.7	24.0	23.2
Grow the Army	6.6	5.4	10.2	8.4	16.2	15.3	19.1	18.0	24.0	23.2

NOTES: Average conditions refers to an unemployment rate of 6.5 percent; favorable conditions refers to an unemployment rate of 8 percent; and unfavorable conditions refers to an unemployment rate of 5 percent. The Grow the Army scenario is based on 80,000 recruits per year and retention rates observed during FYs 2007–2009.

0.5 to 0.7 YOS across all the external conditions we examined relative to the no-wedge results in Table 5.1. This improvement implies that the wedge reduces the loss in average experience levels from roughly 20 percent to roughly 10 percent for grade E-6. This improvement in average experience levels at lower grades is not offset by any loss of experience among higher grades, as E-7, E-8, and E-9 all enjoy relative increases in average YOS with this wedge relative to no wedge.

Table B.3 repeats this exercise for the effects of the regeneration wedge on average YOS under the 980K scenario and should be compared to the results in Table 5.5. It shows a slight improvement in average experience levels across all grades, particularly lower grades.

The results in Tables B.2 and B.3 suggest that, under a five-year time frame to regenerate, use of a wedge—retaining higher-ranking enlisted soldiers and NCOs to serve as leaders under any buildup—does not significantly reduce the risk associated with the lowering of experience likely to occur during a rapid buildup. These results suggest that the possible increases in average NCO experience are slight and would have to be compared against the costs associated with maintaining such a wedge during peacetime.

Table B.3
980K Scenario with Regeneration Wedge: Estimated NCO Experience

	E-5		E-6		E-7		E-8		E-9	
	Steady State	Minimum	Steady State	Minimum	Steady State	Minimum	Steady State	Minimum	Steady State	Minimum
Lesser enlistment eligibility (high-quality mix)										
Average conditions	6.6	5.6	10.2	8.6	16.2	15.7	19.1	18.0	24.1	23.4
Favorable conditions	6.6	5.4	10.2	8.9	16.2	15.7	19.1	18.0	24.1	23.4
Unfavorable conditions	6.6	5.7	10.2	8.7	16.2	15.9	19.1	17.8	24.1	23.4
Greater enlistment eligibility (lower-quality mix)										
Average conditions	6.6	4.8	10.2	9.7	16.2	14.2	19.1	19.0	24.1	23.4
Favorable conditions	6.6	4.8	10.2	9.7	16.2	14.2	19.1	19.0	24.1	23.4
Unfavorable conditions	6.6	5.0	10.2	9.4	16.2	14.7	19.1	17.8	24.1	23.4
Grow the Army	6.6	5.6	10.2	8.6	16.2	15.9	19.1	17.8	24.1	23.4

NOTES: *Average conditions* refers to an unemployment rate of 6.5 percent; *favorable conditions* refers to an unemployment rate of 8 percent; and *unfavorable conditions* refers to an unemployment rate of 5 percent. The Grow the Army scenario is based on 80,000 recruits per year and retention rates observed during FYs 2007–2009.

Abbreviations

AC	active component
ADOS	active duty for operational support
AFQT	Armed Forces Qualification Test
AOS	additional obligated service
AR	Army regulation
ARNG	Army National Guard
BCT	brigade combat team
BLS	U.S. Bureau of Labor Statistics
BOG	boots-on-the-ground
CBO	Congressional Budget Office
CMF	career management field
DEP	delayed entry pool
DoD	U.S. Department of Defense
DMDC	Defense Manpower Data Center
DSG	Defense Strategic Guidance
ETS	expiration of the term of service
FM&C	Financial Management and Comptroller
FY	fiscal year
GAO	Government Accountability Office
IPM	inventory projection model

MBP	monthly basic pay
MOB	mobilization
MOS	military occupational specialty
NCO	noncommissioned officer
OEF	Operation Enduring Freedom
OIF	Operation Iraqi Freedom
OPRA	On-Production Regular Army
OUSD(C)	Office of the Under Secretary of Defense (Comptroller)
RA Analyst	Regular Army Analyst
RC	reserve components
RRF	Required Recruiting Force
SRB	Selective Reenlistment Bonus
SRBM	Selective Reenlistment Bonus multiplier
TAPDB	Total Army Personnel Database
TTHS	transients, trainees, holdees, and students
TRADOC	U.S. Army Training and Doctrine Command
USAR	U.S. Army Reserve
USC	U.S. Code
YOS	years of service

References

"Afghanistan," webpage, Gallup, undated. As of July 29, 2015:
http://www.gallup.com/poll/116233/afghanistan.aspx

AR—*See* Army Regulation.

Army Regulation 525-29, *Army Force Generation*, Washington, D.C.: Headquarters, Department of the Army, 2011.

Army Regulation 600-8-19, *Enlisted Promotions and Reductions*, Washington, D.C.: Department of the Army, 2011.

Army Regulation 601-280, *Army Retention Program*, Washington, D.C.: Headquarters, Department of the Army, 2011.

Arnold, Scott, Christopher Crate, Steven Drennan, Jeffrey Gaylord, Arthur Hoffmann, Donna Martin, Herman Orgeron, and Money Willoughby, *Non-Deployable Soldiers: Understanding the Army's Challenge*, Carlisle Barracks, Pa.: U.S. Army War College, 2011.

Asch, Beth J., Paul Heaton, James Hosek, Francisco Martorell, Curtis Simon, and John T. Warner, *Cash Incentives and Military Enlistment, Attrition, and Reenlistment*, Santa Monica, Calif.: RAND Corporation, MG-950-OSD, 2010. As of September 14, 2015:
http://www.rand.org/pubs/monographs/MG950.html

Asch, Beth J., Michael G. Mattock, and James Hosek, *Reforming Military Retirement: Analysis in Support of the Military Compensation and Retirement Modernization Commission*, Santa Monica, Calif.: RAND Corporation, RR-1022-MCRMC, 2015. As of September 14, 2015:
http://www.rand.org/pubs/research_reports/RR1022.html

Ballard, Tina, Letter to the Honorable Henry A. Waxman, in "Iraqi Reconstruction: Reliance on Private Military Contractors and Status Report," hearing before the Committee on Oversight and Government Reform, 110th Cong., 1st Sess., 2007, pp. 47–48.

Banco, Erin, "Army to Cut Its Forces by 80,000 in 5 Years," *New York Times*, June 25, 2013. As of December 10, 2015:
http://www.nytimes.com/2013/06/26/us/army-to-cut-its-forces-by-80000-in-5-years.html

Bates, Matthew, "Chief of Staff: Warrior Airmen New Culture of Air Force," press release, U.S. Air Force, February 6, 2007. As of May 6, 2016:
http://www.af.mil/News/ArticleDisplay/tabid/223/Article/128105/chief-of-staff-warrior-airmen-new-culture-of-air-force.aspx

Belasco, Amy, *Troop Levels in the Afghan and Iraq Wars, FY2001–FY2012: Cost and Other Potential Issues*, Washington, D.C.: Congressional Research Service, R40682, July 2, 2009.

BLS—*See* U.S. Bureau of Labor Statistics.

Bonds, Timothy M., Dave Baiocchi, and Laurie L. McDonald, *Army Deployments to OIF and OEF*, Santa Monica, Calif.: RAND Corporation, DB-587-A, 2010. As of September 14, 2016: http://www.rand.org/pubs/documented_briefings/DB587.html

Brown, John Sloan, *Kevlar Legions: The Transformation of the U.S. Army, 1989–2005*, Washington, D.C.: Center of Military History, U.S. Army, 2011.

Brownlee, R. L., and Peter J. Schoomaker, *The United States Army 2004 Posture Statement*, Washington, D.C.: Headquarters, Department of the Army, 2004.

Bush, George W., "Ordering the Ready Reserve of the Armed Forces to Active Duty and Delegating Certain Authorities to the Secretary of Defense and the Secretary of Transportation," Executive Order 13223, September 14, 2001, published in *Federal Register*, Vol. 66, No. 181, September 18, 2001, pp. 48201–48202. As of December 10, 2015: https://www.gpo.gov/fdsys/pkg/FR-2001-09-18/pdf/01-23359.pdf

Carlson, Darren K., "Public Support for Military Draft Low," news release, Gallup, November 18, 2003.

———, "Americans Remain Down on Draft," news release, Gallup, November 9, 2004.

Carvalho, Ricardo S., Scott R. Turner, Caitlin Krulikowski, Sean M. Marsh, Andrea B. Zucker, and Matt Boehmer, *Department of Defense Youth Poll Wave 19—June 2010: Overview Report*, Arlington, Va.: Department of Defense, Joint Advertising, Marketing, Research and Studies, December 2010.

CBO—*See* Congressional Budget Office.

Commission on the National Guard and Reserves, *Transforming the National Guard and Reserves into a 21st-Century Operational Force*, Washington, D.C., January 31, 2008. As of December 10, 2015: https://www.loc.gov/rr/frd/pdf-files/CNGR_final-report.pdf

Commission on Wartime Contracting in Iraq and Afghanistan, *Transforming Wartime Contracting: Controlling Costs, Reducing Risks*, 2011. As of December 10, 2015: http://cybercemetery.unt.edu/archive/cwc/20110929213820/http://www.wartimecontracting.gov/docs/CWC_FinalReport-lowres.pdf

Congressional Budget Office, *Recruiting, Retention, and Future Levels of Military Personnel*, Washington, D.C.: Congressional Budget Office, October 2006.

———, *Evaluating Military Compensation Levels*, Washington, D.C., June 29, 2007.

Connable, Ben, and Martin C. Libicki, *How Insurgencies End*, Santa Monica, Calif.: RAND Corporation, MG-965-MCIA, 2010. As of September 14, 2016: http://www.rand.org/pubs/monographs/MG965.html

Cox, Matthew, "Army Has 50,000 Active Soldiers Who Can't Deploy, Top NCO Says," Military.com, November 25, 2015. As of May 2, 2016: http://www.military.com/daily-news/2015/11/25/army-has-50000-active-soldiers-who-cant-deploy-top-nco-says.html

Defense Finance and Accounting Service, "Military Pay Chart, 2015," in Military Pay Charts—1949 to 2016, website, 2016. As of November 3, 2016: http://www.dfas.mil/militarymembers/payentitlements/military-pay-charts.html

Defense Manpower Data Center, "Active Duty Military Personnel by Service by Region/Country," webpage, undated.

———, "U.S. Active Duty Military Deaths, 1980–2010," Defense Casualty Analysis System webpage, November 2011. As of July 30, 2015: https://www.dmdc.osd.mil/dcas/pages/report_by_year_manner.xhtml

———, "U.S. Military Casualties—Operation Enduring Freedom (OEF) Casualty Summary by Month and Service," Defense Casualty Analysis System webpage, 2015a. As of December 10, 2015: https://www.dmdc.osd.mil/dcas/pages/report_oef_month.xhtml

———, "U.S. Military Casualties—Operation Iraqi Freedom (OIF) Casualty Summary by Month and Service," Defense Casualty Analysis System webpage, 2015b. As of December 10, 2015: https://www.dmdc.osd.mil/dcas/pages/report_oif_month.xhtml

Defense Science Board, *Deployment of Members of the National Guard and Reserve in the Global War on Terrorism*, Washington, D.C.: Office of the Under Secretary of Defense for Acquisition, Technology and Logistics, 2007.

Department of Defense Directive 1200.17, "Managing the Reserve Components as an Operational Force," Washington, D.C.: U.S. Department of Defense, October 29, 2008. As of December 10, 2015:
http://www.dtic.mil/whs/directives/corres/pdf/120017p.pdf

———, 1235.10, *Activation, Mobilization, and Demobilization of the Ready Reserve*, November 26, 2008 (Incorporating Change 1, September 21, 2011).

DMDC—*See* Defense Manpower Data Center.

DoD—*See* U.S. Department of Defense.

Donnelly, William M., *Transforming an Army at War: Designing the Modular Force, 1991–2005*, Washington, D.C.: U.S. Army Center of Military History, 2007.

Eshbaugh-Soha, Matthew, and Christopher Linebarger, "Presidential and Media Leadership of Public Opinion on Iraq," *Foreign Policy Analysis*, Vol. 10, No. 4, October 2014, pp. 351–369.

Feickert, Andrew, *Army Drawdown and Restructuring: Background and Issues for Congress*, Washington, D.C., February 28, 2014.

Field Manual 3-0, *Operations*, Washington, D.C.: Headquarters, Department of the Army, 2008.

FM&C—*See* Office of the Assistant Secretary of the Army (Financial Management and Comptroller).

GAO—*See* U.S. Government Accountability Office.

Geren, Pete, and George W. Casey, Jr., *2008 Army Posture Statement*, Washington, D.C.: Headquarters, Department of the Army, 2008.

Grass, Frank, "Authorities and Assumptions Related to Rotational Use of the National Guard," letter to the Chief of Staff, Army, May 31, 2013.

Hamilton, Lee H., and James Baker, *The Iraq Study Group Report*, New York: Vintage Books, 2006.

Hansen, Michael L., and Jennie W. Wenger, *Why Do Pay Elasticity Estimates Differ?* Alexandria, Va.: Center for Naval Analyses, March 2002.

Harvey, Francis J., and Peter J. Schoomaker, *2005 Posture Statement*, Washington, D.C.: Headquarters, Department of the Army, 2005.

———, *2006 Posture Statement: A Campaign Quality Army With Joint And Expeditionary Capabilities*, Washington, D.C.: Headquarters, Department of the Army, 2006.

———, *2007 Posture Statement: A Campaign Quality Army With Joint And Expeditionary Capabilities*, Washington, D.C.: Headquarters, Department of the Army, 2007.

Henning, Charles A., *U.S. Military Stop Loss Program: Key Questions and Answers*, Washington, D.C.: Congressional Research Service, R40121, July 10, 2009.

Horowitz, Stanley A., Robert J. Atwell, and Shaun K. McGee, *Analyzing the Costs of Alternative Army Active/Reserve Force Mixes: Interim Report*, Alexandria, Va.: Institute for Defense Analyses, June 2012.

House Armed Services Committee, *Committee Defense Review Report*, Washington, D.C.: U.S. House of Representatives, December 2006.

"Iraq," webpage, Gallup, undated. As of July 29, 2015:
http://www.gallup.com/poll/1633/iraq.aspx

Johnson, Stuart E., John E. Peters, Karin E. Kitchens, Aaron L. Martin, and Jordan R. Fischbach, *A Review of the Army's Modular Force Structure*, Santa Monica, Calif.: RAND Corporation, TR-972-2-OSD, 2012. As of September 14, 2016:
http://www.rand.org/pubs/technical_reports/TR927-2.html

Joint Publication 1-02, *Department of Defense Dictionary of Military and Associated Terms*, November 8, 2010, as amended through November 15, 2015. As of December 10, 2015:
http://www.dtic.mil/doctrine/new_pubs/jp1_02.pdf

Jones, Jeffrey M., "Vast Majority of Americans Opposed to Reinstituting Draft," news release, Gallup, September 7, 2007.

Kapp, Lawrence, *Recruiting and Retention in the Active Component Military: Are There Problems?* Washington, D.C.: Congressional Research Service, RL31297, February 25, 2002.

———, *Recruiting and Retention: An Overview of FY2005 and FY2006 Results for Active and Reserve Component Enlisted Personnel*, Washington, D.C.: Congressional Research Service, January 20, 2006.

Kapp, Lawrence, and Charles A. Henning, *Recruiting and Retention: An Overview of FY2006 and FY2007 Results for Active and Reserve Component Enlisted Personnel*, Washington, D.C.: Congressional Research Service, February 7, 2008.

———, *Recruiting and Retention: An Overview of FY2008 and FY2009 Results for Active and Reserve Component Enlisted Personnel*, Washington, D.C.: Congressional Research Service, November 30, 2009.

Kliesen, Kevin L., "The 2001 Recession: How Was It Different and What Developments May Have Caused It?" *Federal Reserve Bank of St. Louis Review*, Vol. 85, No. 5, September/October 2003, pp. 23–38.

Klimas, Joshua, Richard E. Darilek, Caroline Baxter, James Dryden, Thomas F. Lippiatt, Laurie L. McDonald, J. Michael Polich, Jerry M. Sollinger, and Stephen Watts, *Assessing the Army's Active-Reserve Component Force Mix*, Santa Monica, Calif.: RAND Corporation, RR-417-1-A, 2014. As of September 14, 2016:
http://www.rand.org/pubs/research_reports/RR417-1.html

Langdon, David S., Terence M. McMenamin, and Thomas J. Krolik, "U.S. Labor Market In 2001: Economy Enters a Recession," *Monthly Labor Review*, February 2002, pp. 3–33.

Lopez, C. Todd, "Odierno: Force Reductions Will Be Responsible, Controlled," U.S. Army website, January 27, 2012. As of May 2015 at:
http://www.army.mil/article/72692/Odierno__Force_reductions_will_be_responsible__controlled/

McHugh, John M., and George W. Casey, Jr., *2010 Army Posture Statement*, Washington, D.C.: Headquarters, Department of the Army, 2010.

McHugh, John M., and Raymond T. Odierno, *2014 Army Posture Statement*, Washington, D.C.: Headquarters, Department of the Army, 2014.

———, *2015 Army Posture Statement*, Washington, D.C.: Headquarters, Department of the Army, 2015.

Odierno, Raymond T., Army Briefing on the FY-13 Budget Request, January 27, 2012.

Office of the Assistant Secretary of the Army (Financial Management and Comptroller), *Department of the Army, FY2005 Budget Estimates, Military Personnel, Army*, Washington, D.C.: Headquarters, Department of the Army, February 2004.

———, *Department of the Army, Fiscal Year (FY) 2006/FY 2007 Budget Estimates, Military Personnel, Army*, Washington, D.C.: Headquarters, Department of the Army, February 2005a.

———, *Department of the Army, Fiscal Year (FY) 2006/FY 2007 Budget Estimates, Reserve Personnel, Army*, Washington, D.C.: Headquarters, Department of the Army, February 2005b.

———, *Department of the Army, Fiscal Year (FY) 2006/FY 2007 Budget Estimates, National Guard Personnel, Army*, Washington, D.C.: Headquarters, Department of the Army, February 2005c.

———, *Department of the Army, Fiscal Year (FY) 2007 Budget Estimates, Military Personnel, Army*, Washington, D.C.: Headquarters, Department of the Army, February 2006a.

———, *Department of the Army, Fiscal Year (FY) 2007 Budget Estimates, National Guard Personnel, Army*, Washington, D.C.: Headquarters, Department of the Army, February 2006b.

———, *Department of the Army, Fiscal Year (FY) 2007 Budget Estimates, Reserve Personnel, Army*, Washington, D.C.: Headquarters, Department of the Army, February 2006c.

———, *Department of the Army, Fiscal Year (FY) 2008/FY 2009 Budget Estimates, Military Personnel, Army*, Washington, D.C.: Headquarters, Department of the Army, February 2007a.

———, *Department of the Army, Fiscal Year (FY) 2008/FY 2009 Budget Estimates, National Guard Personnel, Army*, Washington, D.C.: Headquarters, Department of the Army, February 2007b.

———, *Department of the Army, Fiscal Year (FY) 2008/FY 2009 Budget Estimates, Reserve Personnel, Army*, Washington, D.C.: Headquarters, Department of the Army, February 2007c.

———, *Department of the Army, Fiscal Year (FY) 2010 Budget Estimates, Military Personnel, Army*, Washington, D.C.: Headquarters, Department of the Army, May 2009.

———, *Department of the Army, Fiscal Year (FY) 2011 Budget Estimates, Military Personnel, Army*, Washington, D.C.: Headquarters, Department of the Army, February 2010.

———, *Department of the Army, Fiscal Year (FY) 2012 Budget Estimates, Military Personnel, Army*, Washington, D.C.: Headquarters, Department of the Army, February 2011.

———, *Department of the Army, Fiscal Year (FY) 2013 Budget Estimates, Military Personnel, Army*, Washington, D.C.: Headquarters, Department of the Army, February 2012.

———, *Department of the Army, Fiscal Year (FY) 2013 Budget Estimates, Military Personnel, Army*, Washington, D.C.: Headquarters, Department of the Army, April 2013.

———, *Department of the Army, Fiscal Year (FY) 2016 President's Budget Submission, Military Personnel, Army*, Washington, D.C.: Headquarters, Department of the Army, February 2015a.

———, *Department of the Army, Fiscal Year (FY) 2016 Budget Estimates, National Guard Personnel, Army*, Washington, D.C.: Headquarters, Department of the Army, February 2015b.

———, *Department of the Army, President's Budget 2016, Reserve Personnel, Army*, Washington, D.C.: Headquarters, Department of the Army, February 2015c.

Office of the Under Secretary of Defense (Comptroller), *FY 2003 DoD Agency Financial Report (AFR)/DoDPerformance and Accountability Report (PAR)*, Pt. 5, App. A: *Detailed Performance Metrics*, Washington, D.C.: U.S. Department of Defense, December 23, 2003. As of January 31, 2017: http://comptroller.defense.gov/Financial-Management/Reports/afr2003/

————, *FY 2005 DoD Agency Financial Report (AFR)/DoD Performance and Accountability Report (PAR): Detailed Performance Information for Part 2*, Washington, D.C.: U.S. Department of Defense, November 15, 2005. As of January 31, 2017:
http://comptroller.defense.gov/Financial-Management/Reports/afr2005/f

————, *FY 2006 DoD Agency Financial Report (AFR)/DoDPerformance and Accountability Report (PAR): Detailed Performance Information for Section 2*, Washington, D.C.: U.S. Department of Defense, November 15, 2006. As of January 31, 2017:
http://comptroller.defense.gov/Financial-Management/Reports/afr2006/

————, *National Defense Budget Estimates for FY 2015*, Washington, D.C.: U.S. Department of Defense, April 2014.

Office of the Under Secretary of Defense (Personnel and Readiness), "Military Compensation Background Papers," Washington, D.C.: Department of Defense, May 2005.

Orvis, Bruce R., Steven Garber, Philip Hall-Partyka, Christopher E. Maerzluft, and Tiffany Tsai, *Recruiting Strategies to Support the Army's All-Volunteer Force*, Santa Monica, Calif.: RAND Corporation, RR-1211-A, 2016. As of September 14, 2016:
http://www.rand.org/pubs/research_reports/RR1211.html

OUSD(C)—*See* Office of the Under Secretary of Defense (Comptroller).

Pew Research Center, *The Military-Civilian Gap: War and Sacrifice in the Post-9/11 Era*, Washington, D.C., October 5, 2011.

Public Law 112-25, Budget Control Act of 2011, August 2, 2011.

Public Law 108-136, National Defense Authorization Act for Fiscal Year 2004, November 24, 2003.

Schoomaker, Eric B., Richard A. Stone, Darryl A. Williams, and Brian C. Lein, "Supporting the Deployment of Healthy, Resilient and Fit Soldiers: Soldier Medical Readiness," briefing slides, October 10, 2011.

Schoomaker, Peter J., "Prepared Statement by GEN Peter J. Schoomaker, USA," testimony for the hearing on the Status of the U.S. Army and U.S. Marine Corps in Fighting the Global War on Terrorism, before the Committee on Armed Services, U.S. Senate, 109th Cong., 1st Sess., June 30, 2005, pp. 19–23. As of January 19, 2017:
https://www.gpo.gov/fdsys/pkg/CHRG-109shrg28577/pdf/CHRG-109shrg28577.pdf

Schwartz, Moshe, and Jennifer Church, *Department of Defense's Use of Contractors to Support Military Operations: Background, Analysis, and Issues for Congress*, Washington, D.C.: Congressional Budget Office, R43074, May 17, 2013.

Shanker, Thom, and Elisabeth Bumiller, "In New Strategy, Panetta Plans Even Smaller Army," *New York Times*, January 4, 2012. As of December 10, 2015:
http://www.nytimes.com/2012/01/05/us/in-new-strategy-panetta-plans-even-smaller-army.html?_r=0

Shanker, Thom, and Helene Cooper, "Pentagon Plans to Shrink Army to Pre–World War II Level," *New York Times*, February 23, 2014. As of December 10, 2015:
http://www.nytimes.com/2014/02/24/us/politics/pentagon-plans-to-shrink-army-to-pre-world-war-ii-level.html

Simon, Curtis J., and John T. Warner, "Army Re-enlistment during OIF/OEF: Bonuses, Deployment, and Stop-Loss," *Defense and Peace Economics,* Vol. 21, Nos. 5–6, 2010, pp. 507–527.

Sprenger, Sebastian, "DoD Makes It Official: Budget Cuts Will Shrink Army to 420,000 Soldiers," InsideDefense.com, January 10, 2014.

Tama, Jordan, "The Power and Limitations of Commissions: The Iraq Study Group, Bush, Obama and Congress," *Presidential Studies Quarterly*, Vol. 1, No. 1, March 2007, pp. 135–155.

Tedin, Kent, Brandon Rottinghaus, and Harrell Rodgers, "When the President Goes Public: The Consequences of Communication Mode for Opinion Change Across Issue Types and Groups," *Political Research Quarterly*, Vol. 64, No. 3, September 2011, pp. 506–519.

U.S. Army, "Army Force Mix: Least Risk, Best Value," Army briefing, January 6, 2014.

U.S. Army Medical Command, *Soldier Medical Readiness Campaign Plan 2011–2016*, Vers. 1.2, May 2011.

U.S. Bureau of Labor Statistics, "CPI Inflation Calculator," undated a. As of September 19, 2016:
http://www.bls.gov/data/inflation_calculator.htm

———, *Labor Force Statistics from the Current Population Survey*, webpage, undated b. As of July 27, 2015:
http://www.bls.gov/data/

———, "Household Data Annual Averages, 1. Employment Status of the Civilian Noninstitutional Population, 1945 to date," 2014. As of August 6, 2015:
http://www.bls.gov/cps/cpsaat01.pdf

———, "Consumer Price Index—All Urban Consumers," data series, September 24, 2015. As of September 24, 2015:
http://www.bls.gov/cpi/#data

USC—*See* U.S. Code.

U.S. Code, Title 10, §123, Authority to Suspend Officer Personnel Laws During War or National Emergency.

———, §123a, Suspension of End-Strength and Other Strength Limitations in Time of War or National Emergency.

———, §527, Authorized Strength: General and Flag Officers on Active Duty.

———, §12006, Strength Limitations: Authority to Waive in Time of War or National Emergency.

———, §12301, Reserve Components Generally. As of January 30, 2017:
https://www.gpo.gov/fdsys/pkg/USCODE-2011-title10/pdf/USCODE-2011-title10-subtitleE-partII-chap1209-sec12301.pdf

———, §12302, Ready Reserve. As of December 16, 2015:
https://www.gpo.gov/fdsys/pkg/USCODE-2014-title10/pdf/USCODE-2014-title10-subtitleE-partII-chap1209-sec12302.pdf

———, §12304, Selected Reserve and Certain Individual Ready Reserve Members; Order to Active Duty Other Than During War or National Emergency. As of December 16, 2015:
https://www.gpo.gov/fdsys/pkg/USCODE-2014-title10/pdf/USCODE-2014-title10-subtitleE-partII-chap1209-sec12304.pdf

———, §12304a, Army Reserve, Navy Reserve, Marine Corps Reserve, and Air Force Reserve: Order to Active Duty to Provide Assistance in Response to a Major Disaster or Emergency." As of December 16, 2015:
https://www.gpo.gov/fdsys/pkg/USCODE-2014-title10/pdf/USCODE-2014-title10-subtitleE-partII-chap1209-sec12304a.pdf

———, §12304b, Selected Reserve: Order to Active Duty for Preplanned Missions in Support of the Combatant Commands. As of December 16, 2015:
https://www.gpo.gov/fdsys/pkg/USCODE-2014-title10/pdf/USCODE-2014-title10-subtitleE-partII-chap1209-sec12304b.pdf

———, §12304(f), Notification of Congress.

———, §12305, Authority of President to Suspend Certain Laws Relating to Promotion, Retirement, and Separation. As of December 16, 2015:
https://www.gpo.gov/fdsys/pkg/USCODE-2014-title10/pdf/USCODE-2014-title10-subtitleE-partII-chap1209-sec12305.pdf

U.S. Department of Defense, *Quadrennial Defense Review Report*, Washington, D.C., February 2006.

———, "DoD News Briefing with Secretary Gates and Adm. Mullen from the Pentagon," July 20, 2009.

———, *Sustaining U.S. Global Leadership: Priorities for 21st Century Defense*, Washington, D.C., January 2012.

———, "Statement on Strategic Choices and Management Review," Washington, D.C., July 31, 2013.

U.S. Government Accountability Office, *Military Personnel: DoD Needs to Address Long-Term Reserve Force Availability and Related Mobilization and Demobilization Issues*, Washington, D.C., GAO-04-1031, September 15, 2004.

———, *Reserve Forces: Army Needs to Finalize an Implementation Plan and Funding Strategy for Sustaining an Operational Reserve Force*, Washington, D.C., GAO-09-898, September 2009.

———, *Military Training: Actions Needed to Assess Workforce Requirements and Appropriate Mix of Army Training Personnel*, Washington, D.C., GAO-11-845, September 20, 2011.

U.S. House of Representatives, "The Adequacy of Army Forces," hearing before the Military Personnel Subcommittee of the Committee on Armed Services, 109th Cong., 1st Sess., 2005.

———, "Implications of Iraq Policy on Total Force Readiness," Committee on Armed Services, 110th Cong., 1st Sess., Washington, D.C., January 23, 2007a.

———, "Active Army, Army National Guard, and Army Reserve Recruiting and Retention Programs," hearing before the Military Personnel Subcommittee of the Committee on Armed Services, 110th Cong., 1st Sess., Washington, D.C., August 1, 2007b.

———, "National Defense Authorization Act for Fiscal Year 2011 and Oversight of Previously Authorized Programs," hearing before the Committee on Armed Services, 111th Cong., 2nd Sess., Washington, D.C., March 23, 2010.

U.S. Senate, "Status of the U.S. Army and U.S. Marine Corps in Fighting the Global War on Terrorism," hearing before the Committee on Armed Services, 109th Cong., 1st Sess., Washington, D.C., June 30, 2005.

———, "Nominations Before the Senate Armed Services Committee," 110th Cong., 1st Sess., Washington, D.C.: Headquarters, Department of the Army, 2007.

———, "The Current Status of U.S. Ground Forces," hearing before the Subcommittee on Readiness and Management Support of the Committee on Armed Services, 111th Cong., 1st Sess. 2009.

Wright, Donald P., and Timothy R. Reese, with the Contemporary Operations Study Team, *On Point II: Transition to the New Campaign: The United States Army in Operation IRAQI FREEDOM, May 2003–January 2005*, Ft. Leavenworth, Kan.: Combat Studies Institute, June 2008.